HOW TO DESIGN, BUILD, & TEST COMPLETE SPEAKER SYSTEMS

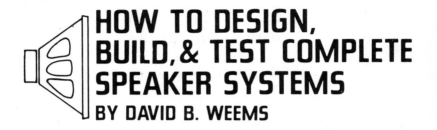

HOW TO DESIGN, BUILD, & TEST COMPLETE SPEAKER SYSTEMS
BY DAVID B. WEEMS

FIRST EDITION

FOURTH PRINTING—JULY 1980

Copyright © 1978 by TAB BOOKS Inc.

Printed in the United States of America

Reproduction or publication of the content in any manner, without express permission of the publisher, is prohibited. No liability is assumed with respect to the use of the information herein.

Library of Congress Cataloging in Publication Data

Weems, David B.
  How to design, build, and test complete speaker systems.

  Includes index.
  1. Loud-speakers—Design and construction—Amateurs' manuals. I. Title.
  TK9968.W43   621.38'0282   78-12710
  ISBN 0-8306-9897-3
  ISBN 0-8306-1064-2 pbk.

Contents

1 Characteristics of Loudspeakers 9
How a Speaker Works—Frequency Response—Dispersion—Speaker Efficiency—Magnets—Power Rating—Cone Resonance—Speaker Impedance—Speaker Q—Transient Response.

2 Multiple Driver Speaker Systems 26
Single vs. Multiple Drivers—Woofer-Tweeter Compatibility—How to Choose a Mid-range Driver—How to Choose Crossover Frequencies to Match Drivers—Crossover Networks—How to Use Piezoelectric Tweeters—How to Make a Crossover Network—How to Design a Simple 6 dB/Octave 2-Way Crossover Network—How to Allow for Voice Coil Inductance—Project 1: A Simple 2-Way System—How to Design a 6 dB/Octave 3-Way Parallel Network—How to Design 12 dB/Octave Crossover Networks—How to Design a 3-Way 12dB/Octave Crossover Network—Project 2: A 3-Way Speaker System—How to Choose Component Values—Final Suggestions.

3 How to Test Speakers and Speaker System Components 57
Useful Test Instrument—Homemade Test Equipment—How to Plot an Impedance Curve—Test Procedures.

4 A Box for Your Speaker 100
Desirable Enclosure Characteristics—Enclosure Materials—Tools and Construction Techniques—How Enclosure Shape Affects Sound—The Fibonacci Series and Enclosure Design—How to Use the "Golden Ratio"—How to Install Speakers—Speaker Placement on Baffle—Damping Material—Grille Cloth—Finishing Your Cabinet—How to Improve a Cheap Commercial Enclosure.

5 Types of Enclosures 125
Closed Box Enclosures—Ported Box Enclosures—Labyrinths and Transmission Lines—Horns—Hybrids—Which Enclosure Is Best?

6 Closed Box Systems 141
How to Fit Box Volume to Speaker—How to Use Test Data to Get the Right Box Size—Project 3: A 3-Way Closed Box Speaker System—Project 4: A 2-Way Closed Box Speaker System—Project 5: A Compact Closed Box Speaker System.

7 Reflex Systems 157
How to Design a Ported Box for a Driver—How to Tune a Ported Box—How to Interpolate From Thiele's Alignment Data—Using a Pocket Calculator to Design Loudspeaker Enclosures—Equalized Reflex System—Designing on Equalized Reflex System: A Worked Example—How to Design a Passive Radiator Reflex—Project 6: Electro-Voice MC8A Ported System—Project 7: JBL Ported System—Project 8: A Compact Ported Box Speaker—The Classic Bass Reflex—Project 9: A Utility Bass Reflex Speaker.

8 **Labyrinths and Transmission Lines** 197
How to Design a Quarter Wavelength Labyrinth—How to Tune a Labyrinth—Transmission Lines—Transmission Line Guides.

9 **Omni-directional Speakers** 206
Kinds of Omni-directional Speakers—Project 10: An Omni-directional Speaker System.

10 **Multiple Speaker Arrays** 213
Enclosures for Arrays—Line Arrays for PA Systems—Rock Concert Line Arrays—Project 11: A Mid-range Array—Project 12: Single Column Stereo System—Project 13: A Small PA Column.

11 **Unusual Speaker Enclosures** 224
Quick Test Speaker—Book End Speakers—Ceramic Tile Enclosures—Project 14: A Ceramic Tile Speaker System—Stuffed Boxes—Project 15: A Compact Stuffed Box Speaker—Single Woofer Stereo—Project 16: Low Cost Sub-woofer.

12 **Contemporary Trends in Speaker Systems** 235
New Kinds of Drivers—New Kinds of Enclosures—Miscellaneous Unconventional Speaker Systems.

13 **Notes on Using Your Speakers** 243
Speaker Placement for Best Frequency Response and Stereo Image—Speaker Position for Good Bass Reproduction—Level and Tone Control Settings—Protecting Your Speakers—How to Hook Up Extension Speakers—How Much Power for Extension Speakers?—Speakers for Quad Systems.

14 **Rainy Day Projects** 253
How to Tame Bad Peaks—Improving Woofers—How to Increase Compliance—How to Increase the Mass of a Cone—Patching Damaged Speakers—What to Do with a Damaged Woofer—Project 17: A Low Cost Woofer-Tweeter System.

15 **Project Plans and Construction Notes** 265
Project 1: A Simple 2-Way System—Project 2: A 3-Way Speaker System—Project 3: A 3-Way Closed Box Speaker System—Project 4: A 2-Way Closed Box Speaker System—Project 5: A Compact Closed Box Speaker System—Project 6: Electro-Voice MC8A Ported System—Project 7: JBL Ported System—Project 8: A Compact Ported Box Speaker—Project 9: A Utility Bass Reflex Speaker—Project 10: An Omni-directional Speaker System—Project 11: A Mid-range Array—Project 12: Single Column Stereo System—Project 13: A Small PA Column—Project 14: A Ceramic Tile Speaker System—Project 15: A Compact Stuffed Box Speaker—Project 16: A Low Cost Woofer-Tweeter System.

16 **On Your Own** 316

Appendix I: Useful Formulas 321

Appendix II: Manufacturers of Raw Frame Loudspeakers 331

Index 335

Introduction

Maybe you have heard the question, "With so many kinds of factory-made speaker systems on the market, why would anybody want to build one?" Anyone who asks that question will probably never understand the answers. One good reason is the same one that makes people tune their own car engine—the fun of making things work right.

If you are a novice at speaker building, welcome aboard! Be prepared to hear some discouraging words when you start to work. Skeptics will say that homemade speakers always look homemade, that you don't have the know-how to do the job right (not in those words, of course), and that it will cost you just as much as a factory system anyway. Some of your friends will probably repeat what I call the Titanic Fallacy.

If remarks about homemade furniture disturb you, let them be a challenge to produce a good looking cabinet. At least it can be unique.

The people who doubt your ability to do something out of the ordinary are really trying to put you into mold. Don't let them. The audio quality of your speakers will depend as much on the attention you give to each detail and on your patience, as on your knowledge of design principles. When you have built your own system, you won't be afraid to tinker with it until it sounds right.

Of course there is always the possibility that you won't get any better sound for the money, but you can aim for the kind of sound or appearance that you want. Representatives of more than one company have admitted to me that the home builder is in a unique position when it comes to using woofers that require large enclosures for optimum performance. The manufacturer must produce speaker systems that appeal to a mass audience; you have only yourself and perhaps your family to please.

Be on guard against the Titanic Fallacy. Its theme is "Trust the equipment." Never mind that there are icebergs out there; the ship is infallible. If an expensive factory-made speaker·system doesn't sound right to you, your ears are wrong. Nonsense. Don't be intimidated by a piece of equipment. The Titanic Fallacy is responsible for all kinds of malfunctioning machines, from toys to automobiles. The curse of this fallacy is that if you don't like something that is touted as the newest and greatest, you are led to doubt your own ability to know what is good, or even what you want. In choosing your own sound system you are king and don't you forget it.

In this book you will find a variety of speaker system projects from simple speaker-in-a-box set ups to more complicated systems using several drivers. You will also find complete building plans for each of them. But, more important, you will find practical information on how to design your own speaker system. The design information is based on theory—but on theory that has been put to test.

I recognized the importance of giving design information recently when my neighbor went to a local high school wood shop to look at a floor model belt sander. He told the shop teacher that he wanted to build one for his home workship. The teacher, trying to be helpful, said, "I've got a set of plans for a homemade sander." My neighbor shook his head. "Don't want any plans. Half the fun is working out the plans." Or, as millions of kids through the ages have said it, "Please, Mother, I want to do it myself." So, if after reading this book you find a project that appeals to you, but you want to tamper with its design—good.

This book owes too much to too many people for me to name them all. We are all indebted to Neville Thiele and Richard Small for their work on enclosure design. I especially appreciate the past help I've received from James F. Novak, Vice President, Engineering, Jensen Sound Laboratories; and for permission to use material from D.B. Keele, Jr., formerly of Electro-Voice, now Chief Engineer of Klipsch & Associates; and from Pat Snyder of Speakerlab. I also owe much to Al Kaler for his help, to Oliver P. Ferrell for past encouragement, and finally to Charys for much help and, even more, for putting up with speaker systems that have crowded every room of our house for longer than I want to remember.

David B. Weems

Characteristics of Loudspeakers

The speaker is the part of a sound system that has the fewest specifications, but the greatest effect on quality. If you switch amplifiers, you may not notice the difference unless you are a critical listener or unless one of the amplifiers has a much greater power output, but a change of speakers can make a striking difference. To up-grade a stereo or quad set, the speaker system is a good place to start.

Anyone who had never seen an audio system—say a Martian—would probably think the speaker is the simplest part. In some ways that's right, but the speaker is also probably the most misunderstood component. For generations it's been called the "weakest link in the audio chain" by both audio engineers and by people who know little about sound equipment, yet blame every shortcoming of their set on the speakers, saying, "It's in the speakers; I can hear it."

Speakers don't have the flat response of tuners, amplifiers, or even of microphones or headphones; but they may be underrated. Consider the speaker's job: it must convert to mechanical energy the electrical energy which it receives from the amplifier. Electric motors do that too—but a speaker is expected to change current alternating at frequencies of 20 to 20,000 times per second into sound waves without altering the signal. Even if the demand on a single cone speaker is relaxed to 50 to 10,000 Hz, the cone may travel as far as ¼ in. and back 50 times per second to produce a 50 Hz tone while it is vibrating at rates up to 10,000 times per second, undergoing constant acceleration first in one direction, then the other. Awesome.

The stresses on the cone defy analysis, but the musical world compounds the speaker's task by its choice of instruments: hollow containers with large air volumes covered with stretched hides (drums); strips of catgut attached to resonant wooden boxes (strings); and metallic pipes that end in flaring horns (brasses), or in tuned resonant air columns up to 32 feet long (organ). Each kind of instrument produces a quality of sound peculiar to its design, and we expect the speaker to reproduce the tone color of each of these. Even more, we except the speaker to combine that of all of them into a composite sound pattern while yet preserving the thread of each so we can hear it in its relation to the whole.

Except for a small minority of specialized speakers, all of these physical and musical demands are put on a small piece of thin paper—the speaker cone of the dynamic speaker. Considering the work cut out for it, the speaker may be the strongest link in the audio chain, even if it is less than perfect.

HOW A SPEAKER WORKS

The working parts of a dymanic speaker are the cone and its suspension, the voice coil, and the magnet (Fig. 1-1). When an alternating current goes through the coil, it produces a magnetic field around the coil that builds up and collapses in response to the frequency of the current. The field in the coil interacts with the

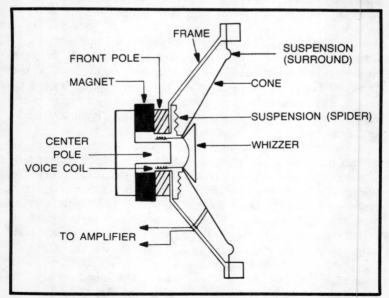

Fig. 1-1. Parts of a loudspeaker.

permanent field of the magnet to apply force to the coil. This force varies in strength according to the current going through the coil, the length of the wire in the coil, and the intensity of the field produced by the speaker's magnet in the gap where the coil is located. When the force is applied, the coil moves. Its movement is controlled by the speaker's suspension system, made up of an outer cone surround and a coil centering device, called the *spider*. These two suspension parts allow the coil to move along the axis of the magnet's center pole, but they prevent it from shifting far enough laterally for the voice coil to rub the pole. The suspension should be designed to permit the cone movement to be linear with respect to the intensity of the signal up to the power limits of the speaker.

An electric motor can be described as "a coil of wire that moves in a magnetic field when a current is applied to it." Speaker engineers often refer to the driving mechanism of a dynamic speaker as a "motor."

If you connect the leads of a sensitive ac voltmeter to the terminals of a dynamic speaker and gently push the cone with your hand, the meter pointer will move—indicating that you have produced a small voltage across the terminals. When mechanical energy is applied to a motor, it can become a generator. Accordingly, speakers connected to the input of high gain amplifiers can act as microphones, changing sound energy to electrical energy. Many intercom sets use this principle. They contain small speakers that work in a dual role: converting electrical energy to sound when the set is operated in the receive mode, or converting sound energy into electrical energy when the set is operated in the sending mode.

When a current passes through the voice coil, the coil moves. When the coil moves, it produces a voltage. The direction of the electromotive force (emf) that is produced is always the one that will oppose the change of current direction in the alternating current that drives the speaker. Because it is in opposition to the driving voltage, this secondary voltage is called the "back emf." The stronger the magnetic field, the greater the back emf and the higher is the electrical damping that controls the voice coil and, thus, the cone action.

FREQUENCY RESPONSE

This is one of the most talked about characteristics of speakers. Stores that sell low-priced speakers often advertise frequency response. The description of such a speaker may read, impedance, 8 Ω; resonance, 60 Hz; magnet, 10 oz.; response, 30 to 18,000 Hz. When the frequency range is stated like this, with no reference made to the limit of variation in output over the range, it means nothing. In this case, you can get an idea of how optimistic the figures are by noting that the free air resonance of the speaker is an octave above the stated low frequency cut-off. Speakers in conventional boxes have a rapid roll-off in low frequency response below their bass resonance frequency. And if the speaker is installed in a compact closed box, that frequency will be raised substantially.

Even when frequency information is more specific, such as 30 to 18,000 Hz and ± 3 dB, it isn't worth much unless the conditions of measurement are stated. If a speaker is measured in a normal living room and then moved a few feet, its second measured response will be significantly different. That is why speaker manufacturers use a "dead" room (a "dead" room is a large room with fiberglass wedges protruding from all floor, ceiling, and wall surfaces to absorb sound) or outdoor measurements for speaker evaluation. By removing the variables of room reflections, design engineers can make more consistent comparisons between speakers and study the effects of design changes.

In selecting a speaker for music reproduction, smoothness of response is worth more than extended range alone. A speaker with a smooth response from 100 to 10,000 Hz can produce music more faithfully than one with a 50 to 15,000 Hz range that has significant peaks in that range. Figure 1-2 shows the frequency response of two speakers from 1,000 to 10,000 Hz. The 4 × 6 in. oval speaker has a smoother high frequency response, although the round 5 in. speaker has a larger magnet and a roll-edge suspension that gives it a lower bass response. At first impression most listeners might prefer the wider range of the 5 in. speaker, but its greater coloration is tiring. The relatively good performance of the cheaper speaker shows an advantage—at least in cheap speakers—for the oval cone shape that has a wider spread of resonances and path lengths. Round cones can be made to have good performance, but they require more careful design.

The requirements for a wide frequency range from a single cone speaker are contradictory. For good performance above 10,000 Hz, the cone must be light, having a mass no greater than about 5 grams. But light cones bend at low frequencies, producing distortion. For good bass response the cone should be large in area so its can "grab" enough air to have good radiation resistance and to be able to pump out bass without excessive movement. If a large cone is reinforced enough to give good piston action at low frequencies, the added mass will lower its frequency of resonance. The bass range will be extended further, but the heavier cone will require more force to start and stop it. This reduces efficiency, and too much cone mass can give a poor transient response, causing the speaker

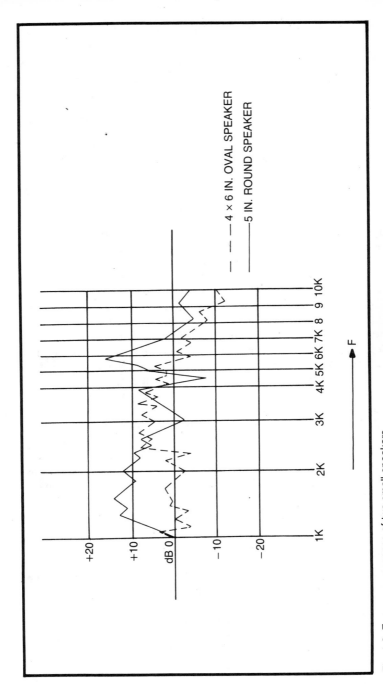

Fig. 1-2. Frequency response of two small speakers.

to blur attacks and have hangover after the signal from the amplifier has ended. The heavier cone often has a smoother frequency response than a light cone of similar size if the mass is added with that purpose in mind, but the heavier one will lack upper highs.

One common method that designers have used to make the most of large cones is to add a secondary smaller one, called a *whizzer*. This whizzer cone is less expensive than a separate tweeter because it is driven by the same magnet and voice coil as the main cone. A whizzer can improve the radiation pattern of the highs as well as extend the frequency range. Its free edge is undamped, which is a theoretical disadvantage, but the whizzer does improve the frequency response of a single cone speaker. It is used on many 8 to 12 in. full range speakers.

Even those large cones that have no whizzer produce some high frequency sound. They do this by cone *reduction*. At low frequencies the entire cone vibrates as a single unit, but as the frequency of the signal is increased, the central part of the cone vibrates almost as a separate unit. This small part of the cone has a lower mass than the whole cone, so it is able to vibrate at relatively high frequencies. Some manufacturers have produced large single cone speakers with decoupling compliances in the cone to encourage the central section to vibrate independently of the outer section. Because of the current emphasis on separate speakers for low and high frequencies, this kind of design is not seen as often today as it was in earlier generations of high fidelity speakers.

Although the outer suspension of a full range speaker is designed mainly to attain the desired compliance and cone resonance, it has another important role—to damp wave motion that travels out from the center of the cone. If higher frequency waves are reflected, they meet new waves coming out and cause interference. Most manufacturers damp the cone rim, either with a foam surround or by adding a viscous material to the surround. Some speakers have a ridge of foam material on the edge of the cone itself for the same purpose. If done properly, such damping material makes for a smoother frequency response.

DISPERSION

Some speakers that have a wide frequency response when measured with a microphone directly in front of the speaker change their character dramatically when you move across their projected sound beam. If you sit directly in front of such a speaker, you will hear wide range sound, but the highs may be harsh because the high frequency energy is concentrated in a narrow pattern. As you move about between two such speakers, the stereo image may shift unnaturally. Before the advent of stereo, people often complained that this kind of speaker produced music that sounded "as if it was coming through a hole in the wall."

During the last decade or so, there has been an increased awareness of the importance of good dispersion. A speaker that spreads out the highs will enlarge the apparent source, quite unlike the keyhole effect. If the good dispersion goes up high enough in frequency, it adds an airy quality to the sound that is similar to that of live music.

Although good dispersion makes a speaker sound big, it really requires a *small* cone. A driver can radiate sound in all directions only if the effective part of the cone action for a given frequency is no larger in diameter than a good fraction of the wavelength of the sound. One rule of thumb is to consider a cone sufficiently omnidirectional up to the point where the effective diameter of the cone equals one wavelength. Following this rule, a 12 in. speaker should not be expected to radiate sound above 1300 Hz, an 8 in. speaker above 2000 Hz or a 4 in. speaker above 4000 Hz. For frequencies above 10,000 Hz, you would need a 1 in. tweeter, and so on. This is not an iron rule for several reasons. Most speakers used for frequencies higher than those mentioned above have either a whizzer or a central dome radiator, or lacking those, they probably have a certain degree of decoupling at high frequencies. If they didn't, all full range speakers would sound terrible.

Some speakers are designed to have better dispersion in the horizontal plane where it is needed than in the vertical plane where the sound mainly hits the carpet or the ceiling. Oval speakers give better dispersion when they are oriented with their long dimension in a vertical line because they then have the width of a round speaker with a smaller cone area. Line radiators (speakers installed in a tall, narrow column), have good horizontal dispersion equal to that of one of the single speakers used in the column. Massed multiple speakers on a flat board have poor radiation patterns because the effective source is much larger; the dispersion characteristics approach those of a very large cone, but with less predicable variations.

One evidence of the recognition of the need for good dispersion is the shrinking of tweeter size from the 3 to 5 in. cones of earlier years to 2 in. or even 1 in. dome tweeters. Manufacturers have developed the dome tweeter because the dome shape provides the necessary strength in the small cone, not because the dome gives better dispersion than other shapes. The size of the cone rather than its shape controls the dispersion characteristics. Awareness of this fact has made some manufacturers reduce the size of their tweeter cones until they are actually smaller than their advertised diameter.

The pay-off of this trend to smaller tweeters is better dispersion and more natural sound.

SPEAKER EFFICIENCY

If you test several speakers by switching from one to another without changing the volume control on your amplifier, you will notice that some produce more sound than others. The measure of a speaker's ability to convert electrical energy into sound energy is its *efficiency*. Mathematically, efficiency is sound output power (in acoustical watts) divided by electrical input power (in electrical watts).

There are two ways to increase the efficiency of a speaker: reducing the cone's resistance to motion, or increasing the force on the voice coil for a given flow of electrical current.

The easiest way to reduce the cone's resistance to motion is to make it lighter. By reducing cone mass to half the original value, the efficiency is increased by four times in the mid frequency band. But light cones usually have a rougher response than heavy cones. Another way to reduce motion resistance is to increase the compliance of the suspension. There is a practical limit here in that, when driven hard, too little stiffness can permit the voice coil to move laterally against the pole piece or bottom.

To get more force, the manufacturer can use a stronger magnet, up to a point, or he can put more turns of wire in the coil. Too much magnet will overdamp the speaker in a closed box, cutting off its bass range at a needlessly high frequency. And using a bigger magnet is expensive. A manufacturer can get more strength from a small magnet by decreasing the width of the gap around the voice coil, but at the risk of a rubbing coil.

The number of turns of wire that can be used is also limited by the available space in the gap, and by there being a point after which the voice coil mass exceeds the optimum ratio of voice coil mass to total moving mass.

For any speaker, the final efficiency is usually a compromise between the demands of physical laws and practical considerations. High efficiency is good because it can save you the price of a higher powered amplifier, but the only aspect of sound that it enhances is dynamic range. And the size and kind of enclosure helps to set the final efficiency of the system.

The measurement of speaker efficiency is a complicated business, as is any kind of dependable acoustical measurement. One of the useful by-products of recent work by Thiele and Small was the discovery that a speaker's reference efficiency can be determined from simple electrical measurements (Test 8, Chapter 3).

Producing a high sound intensity in the average room requires very little acoustical power, which is a good thing since most speakers are inefficient. A room that measures 15 × 25 ft. with an 8 ft. ceiling would require only about 0.0005W of acoustical power to produce an 80 dB sound pressure level, a volume of sound that many people would call "loud." Other listeners might prefer to put the sound and feel of a full symphony orchestra or a live rock concert in such a room. These addicts would want a peak level of at least 110 dB, 30 dB louder than loud. In terms of the acoustical power output (0.5W), that is a thousand times as great. The logarithmic dB scale tells us how much our hearing compresses the sound volume range. We sense a 10 dB change as half or double the volume instead of as the ten-fold difference it really is.

Most speaker manufacturers don't publish efficiency figures for their products, and popular speaker systems can have an efficiency ratio of 20 to 1. That doesn't include large horns, which are by far the most efficient of all. Compact, closed box speakers are the most power-hungry kind, with typical efficiencies of from 0.25% to 0.5%. Early high fidelity speakers in large enclosures had ratings about 10 times greater than today's compacts, and for large horns the figures go to 30% and higher.

Assuming 0.5% efficiency, a speaker would require 100W of electrical power to produce the 110 dB level in the room mentioned above. This would leave no margin for unusual peaks when the volume control is turned a bit higher than usual. But if the listeners were satisfied with a sound level of 80 dB, which is about twice the apparent loudness of enthusiastic conversation, 0.1W from the amplifier would be enough. Obviously, the required amplifier power and speaker efficiency depends more on the listener than on the room. Anyone can cut his amplifier costs by choosing an efficient speaker that is designed to give optimum performance in a medium to large reflex enclosure.

MAGNETS

The magnet of a speaker should be designed to give satisfactory efficiency and to produce the right degree of damping. The first high quality magnets were developed about World War II from alloys that contained aluminum, nickel, and cobalt. The most famous of those was Alnico V, used in many early hi-fi speakers. More recently another kind of magnet, the *ceramic* magnet, has come into general use. Made from strontium ferrite or barium ferrite, ceramic magnets, give the speaker a shallower profile, cost less than Alnico V magnets, and are flatter.

Can you judge the quality of a speaker by magnet weight alone? More important than the magnet's weight is its strength, or flux density, in the magnet gap. The other speaker parts are important too. A cone should respond to the voice coil's demand without adding distortion, and even the frame contributes to a speaker's performance by holding the parts in proper alignment during the life of the speaker. But, because the magnet is the most expensive part, manufacturers rarely squander the cost of a big magnet on a poor speaker. Magnet weight can be a rough guide to speaker quality, but it's not everything. Figure 1-2 proves that.

Here are some typical ceramic magnet weights for various rated speaker sizes: an 8 in. speaker should probably have a magnet weight from 8 oz. to 10 oz.; a 10 in. speaker from 10 oz. to 16 oz.; and a 12 in. model, from 20 oz. to 30 oz. or more. Large, heavy cones need more magnet.

POWER RATING

Sometimes people who buy a speaker with a high power rating are disappointed by the volume of sound they get out of it. A high power rating tells you how much power the speaker can *absorb*—NOT how much sound it will put out. Inefficient speakers usually do have a higher power rating than more efficient speakers, so a speaker with a high power rating can sometimes absorb all the power your amplifier can deliver without producing the sound levels you had anticipated. Power rating without any kind of efficiency rating means nothing.

The power rating that manufacturers assign to speakers is usually the limit of electrical power that each speaker can absorb without damage to the voice coil, or to the cone and its suspension. One reason that speakers lose efficiency is that some of the electrical energy delivered to them is converted into heat by the voice coil's resistance. Too much heat can loosen or warp the voice coil unless it is radiated or conducted away from the coil. The larger the diameter of the voice coil, the better it can dissipate the heat. You can estimate the power handling ability of a speaker by checking the diameter of the voice coil. Speakers with the smallest voice coils, from ½ in. to 9/16 in. in diameter, are usually rated at no more than 5 W. The power ratings gradually rise from that value with increased diameter coils up to speakers with 1 in. voice coils that can usually handle 15 to 30W of power. Voice coils with a diameter that is greater than 1 in. can usually handle much more power. A 2 in. voice coil speaker can be rated at 100W or more.

These power ratings are rms values, which means that you can use an amplifier with a much higher power rating than the speaker.

Music is full of transient sound rather than sustained tones that would drive the speaker harder.

It would be useful to have another kind of power rating, the power level at which the speaker produces a certain level of distortion. Because speakers usually produce much greater distortion than amplifiers, particularly at low frequencies, this kind of power rating is almost never mentioned. To keep distortion low, you should always consider the power rating of your speaker to be somewhat less than its advertised power rating.

CONE RESONANCE

If you suspend a mass on a spring (Fig. 1-3), you can set the mass in motion and it will vibrate at a certain frequency. If you add a blob of modeling clay or any other weight to the mass, you will see the mass vibrate more slowly. Substituting a more compliant spring will also reduce the frequency of the resonance. To increase the frequency, you could either remove part of the mass, or you could use a stiffer spring.

A loudspeaker cone and its suspension behave in the same manner as the mass on the spring. Each speaker's cone resonates at a certain frequency which is determined by the mass of the cone and the compliance of its suspension. Large cones, having greater mass, usually have a lower frequency of resonance than small cones. Resonance measured with the bare speaker operating in free air is called the *free air resonance*.

If you sprinkle some talcum powder on a speaker cone and watch the talcum as you vary the frequency of the drive signal from an audio generator, you will see that the cone's vibration increases

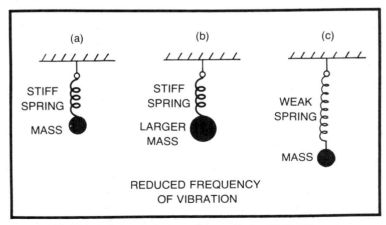

Fig. 1-3. How changing mass or compliance affects resonance.

as you approach the speaker's resonance. Then, at resonance itself, the speaker will vibrate wildly.

At resonance, the speaker is extremely efficient at converting electrical energy to sound. If the magnetic field is weak, as it is in most cheap speakers, the speaker may produce a loud boom whenever the signal from the amplifier contains a tone that is near the speaker's resonance. But because the voice coil acts as a generator as well as a motor, the back emf it produces resists the passage of current through the coil. This back emf peaks when coil motion peaks. The stronger the magnet, the higher the opposition, or *impedance*, to the flow of current; therefore, a strong magnet damps the tendency of the cone to go out of control at resonance. An amplifier with low internal resistance, as many modern amplifiers have, acts like a short across the voice coil, damping cone motion at resonance.

A speaker that is designed for use in a compact sealed box must have a low resonance frequency because the air pressure in the box resists cone movement and, in effect, adds stiffness to the cone. To reduce the frequency of the system resonance, the designers of woofers for closed box operation use floppy suspensions and heavy cones. The traditional accordion suspension, which was a series of corrugations in the outer cone, has been largely replaced by the roll edge. The roll edge is really a half roll, usually with a convex surface, but sometimes with a concave surface if it is being viewed from the front of the speaker. Some speaker engineers claim that the roll edge withstands the pressure differences between the interior and exterior of small boxes better than the accordian edge does.

Various materials such as cloth treated with a sealant, foam plastic, and neoprene are used for the surround. The composition of surround material has evolved; speaker manufacturers have learned which materials decompose or become stiff with age. A good surround material will damp the cone edge and retain its degree of flexibility through changes of temperature and humidity, and will not lose these qualities as it ages. It should not permit air to leak through it, and it should not deform easily. It must permit the cone to travel linearly with the increasing intensity of the signal up to the limit of cone excursion.

SPEAKER IMPEDANCE

When you buy a speaker, you will choose a certain impedance rating to match the speaker to your amplifier. That rating, which the manufacturer puts on the speaker itself or on its specification sheet, will have an average value of 8Ω, more or less, ranging from 4 to 16Ω. This rating suggests that a speaker is like a resistor of the same value: the units (ohms) are the same, but there are important differences between the impedance of a speaker and pure resistance.

If you measure the resistance of your speaker's voice coil with an ohmmeter, you will find that it is about 75% to 80% of the rated impedance. This tells you that impedance is something more than simple resistance. The ohmmeter measures resistance by putting a *direct* current through the voice coil, but the speaker must operate on *alternating* current. When an alternating current passes through a coil, the constantly swinging flow sets up its own magnetic field which grows and collapses with the frequency of the current, moving lines of force through the coil. This causes a reactance in the coil to alternating current; the more rapidly the current reverses itself, the greater the reactance.

In addition to the inductive reactance mentioned above that discriminates against high frequency current, a coil also has some capacitive reactance whose magnitude also varies with frequency. This means that while your speaker may have a certain resistance to a steady dc current, it will have a still greater reactance against an ac current except at frequencies where the capacitive reactance cancels the inductive reactance. The complex quantity of its total opposition to an ac current is its impedance.

The impedance rating for a typical speaker is probably $8\,\Omega$, so you can use that figure to calculate the total impedance if you connect more than one speaker to the same output taps on your amplifier. However, remember that the reactance of the voice coil changes with the rate at which the alternating current reverses itself. That means that the voice coil's impedance varies according to the frequency of the current and will usually oppose high frequencies more than low frequencies because of the voice coil inductance.

If you measure the speaker's impedance at various frequencies and plot a graph of the values you get, the curve will look similar to that in Fig. 1-4. The gradual rise in impedance at high frequencies is no surprise, but why the hump at 50 Hz? This hump corresponds to the frequency of resonance. At resonance, the vibration of the voice coil in the magnetic field increases the back emf, giving the same effect as a rise in impedance.

While a high peak in the impedance curve at resonance may look bad, remember that this is not a response curve. That high peak is a sign of a strong magnetic field, a sign of quality. A smaller magnet would produce a weaker field, reducing the motional back emf, and lowering the impedance peak. In that case, the sound output at resonance would increase because the lower impedance would be permitting increased current. Any acoustical resistance—such as a cloth, or sheet of fiber glass insulation material, stretched tightly over the back of the speaker—will lower the impedance at resonance, but it will also affect the shape of the impedance curve and reduce the cone's efficiency at that frequency.

The rated impedance is the measured impedance at a given frequency chosen above the free air resonance, from between 100 to 1000 Hz (often 400 Hz), which is the region of the frequency band containing the greatest distribution of power in ordinary music or speech. This is also near the low point on the impedance curve for most speakers, so there is no danger of blowing fuses or throwing circuit breakers in your amplifier if you use the rated impedance to maintain the proper minimum impedance load.

One trick used by some manufacturers is to make the real impedance lower than the rated impedance: a speaker sold as an 8Ω speaker may really be a 4Ω speaker. The reason for this ploy is that the lower impedance permits the speaker to draw more current, giving it the appearance of higher efficiency than it really has. If the speaker is connected in parallel with other speakers, this can cause a problem, and is a good reason for testing your own speakers.

SPEAKER Q

The letter "Q" is sometimes used to denote the directivity of a speaker, but more often it describes the characteristics of the speaker's response near resonance. The speaker's acoustical behavior at resonance is similar to the electrical behavior of a resonant electrical circuit, so the electrical term "Q" is applied to the speaker's response curve. In a resonant electrical circuit, the higher the Q, the steeper the skirts of the curve at resonance. The value of Q is equal to the ratio of reactance to resistance in a series circuit or of resistance to reactance in a parallel circuit.

While a speaker's impedance curve looks like the curve in Fig. 1-4, it is the acoustical output near resonance that is important. However, the approximate value of Q can be found from the impedance curve by getting the ratio of the impedance peak (Z) to the voice coil dc resistance (R_e) as well as the shape of the curve. The speaker's total Q is a combination of its mechanical Q and its electrical Q, and the system Q is influenced by the amplifier's damping factor. Typical values for the mechanical Q of speakers range from 2 or 3 up to 12 or higher. Electrical Q varies from a low of about 0.2 up to 2 or 3. A speaker's total Q is somewhat lower than its electrical Q.

A formula for Q, adapted from that shown in *Radiotron Designer's Handbook*, a standard reference, is:

$$Q = f_s/\Delta f \times \frac{R_e + R_g}{Z_{max} + R_g}$$

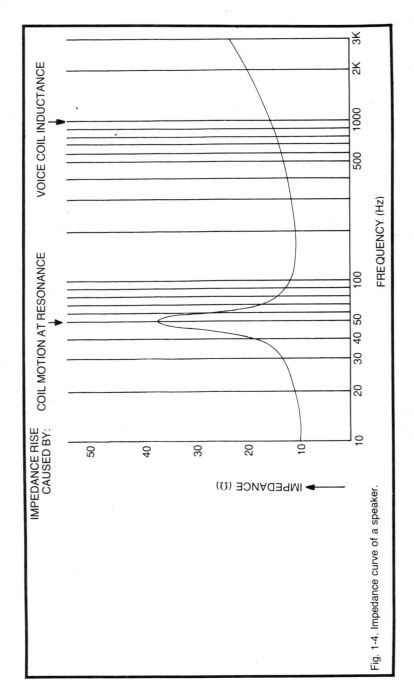

Fig. 1-4. Impedance curve of a speaker.

where f_s = frequency of resonance
Δ_f = band-width at half power point (frequency band-width where power at each end is 3 dB below maximum)
R_e = voice coil resistance
Z_{max} = maximum value of impedance (at f_s)
R_g = output resistance of amplifier and speaker cable

Notice that Q is decreased by a wide band-width between the half power points, which means a broader peak; and by a high impedance at f_s. The effect of R_g is less obvious, but since R_e is always smaller than Z_{max}, a high value for R_g will affect the numerator of the equation more than the demoninator, raising the total Q. Modern amplifiers have such low internal resistance (high damping factor) that R_g usually approaches zero and can be disregarded as a significant contributor to higher Q. Although the formula above can be used to approximate the value of Q by using electrical measurements, a more refined test has been developed by Thiele and Small and is shown in Chapter 3.

It has long been accepted that a speaker in a closed box should have a Q of about 1 for the most extended bass range without excessive peaking at resonance. Speakers with higher values of Q will be subject to boom and hangover, the bad habit of vibrating after the signal ends. A low Q is the earmark of a speaker with a large magnet, but it could also mean that the speaker is overdamped and that its bass response will roll off at a higher point in the frequency scale. When you know the Q of a speaker in a closed box, you can calculate the response at resonance compared to its mid-band response by this formula:

$$\text{Response at resonance} = 20 \log_{10} Q$$

Using this formula, you can look up the log of various values of Q. The results for some sample values of Q are listed in Fig. 1-5. For a speaker with a Q of 2, the output at resonance is 16 times what it would be for a Q of 0.5. The Q should suit the intended use. A speaker with a Q of 2 would have a muddy bass because of hangover. One with a Q of 0.5 would be fine for a speaker designed for use in a reflex enclosure, but in a closed box it would have thin bass unless it were assisted by a special low frequency bass boost circuit.

The most effective way for a manufacturer to get the Q he wants is for him to choose the proper size of magnet. A large magnet decreases Q by damping the cone at resonance.

You can control the Q of your speaker by considering it in the design of your enclosures and in the choice and use of damping material in the enclosure. In fact, the Q of your speaker is one of the most important characteristics to be considered when you plan your system. Small boxes raise Q, but damping material can lower it,

Q	log Q	$20 \log^{10} Q$ (Speaker response at resonance)
0.5	0.6990-1 (-.3010)	−6 dB
1.0	0.00	0 dB
2.0	.3010	+6 dB

Fig. 1-5. Sample values of Q.

particularly when the damping is applied in a stretched collar over the back of the speaker. Modern amplifiers which have a low internal resistance and a high damping factor preserve the low Q that is built into large magnet speakers, but any resistance in your speaker circuit can raise the Q.

Although Q is a characteristic often ignored by an amateur builder of speaker systems, you have a powerful tool for manipulating the bass response of your speaker by controlling its Q.

TRANSIENT RESPONSE

If a speaker has good transient response, it will start to move almost immediately after receiving the amplifier's signal to do so, and it will stop promptly when the signal ends. A good way to judge transient response is to listen carefully to the staccato sound a speaker reproduces, such as the sound of percussion instruments. With a poor speaker, staccato will be mushy and blurred.

A smooth frequency response is the first requirement for good transient response. A peaky response curve indicates multiple cone resonances, and each resonance can be kicked off by any sound containing the resonance frequency in the fundamental tone or its natural overtones. Resonances are hard to control, so every resonance adds its share of hangover, and the aural effect of hangover is a loss of definition and clarity.

Even if a speaker has a smooth frequency response without higher resonances, there is at least one resonance in every speaker system—the bass resonance. This resonance varies with system Q; a high Q speaker has poor magnetic damping. One of the arguments for using the smallest woofer that will be adequate is that small, light cones can be more easily controlled than large, heavy cones. Heavy cones, on the other hand, usually have a smoother frequency response. A confusing factor is the variable sensitivity of the ear to hangover at different frequencies. If the hangover occurs at a very low frequency, it has the effect of pulse stretching, which can make the bass more audible without noticeably changing the character of the overall sound. However, an ideal speaker should reproduce the input signal without any additional hangover at any frequency.

Multiple Driver Speaker Systems

"How many speakers?" is one of the first questions to be faced in designing a speaker system. A visit to any audio dealer will show that manufacturers do not agree on the answer. Different engineers appear to have solved the problems of a full frequency response in different ways—or maybe different engineers solve different problems. If you compare the number of drivers against prices, you will notice that higher priced speakers usually have more complicated crossover networks and more drivers to handle the full audio range. Are these added components necessary? Or are they put in because people who pay more expect more complex systems?

SINGLE VS. MULTIPLE DRIVERS

To do its job, a single cone full range speaker needs a dual personality. It seems part of the natural order to associate good bass response with a large cone and good treble with a small one. Some manufacturers carry the process further by saying, in effect, "Two cones are good; three cones are better." Although there is general agreement that you can get a wider frequency response and lower distortion with multiple drivers, the minority viewpoint must be considered that says, "Two cones are good; one cone is better."

A young woman who knew nothing about sound reproduction but was musically inclined startled some audio fans into attention a few years ago by saying that a full range single cone speaker sounded more natural than several more expensive commercial 2-way and 3-way systems. She was undecided until she heard a vocal recording; then she pointed to the box with an 8 in. full range speaker and said, "That one has a more integrated sound."

Current multiple speaker systems seem to have better integration of sound than earlier models had, but separate source effects are always a danger for a multiple driver system. A more subtle problem is phase distortion, which can be introduced in two ways: by the electrical effects of a crossover network, or by the acoustical shift of phase caused by the spacing between the drivers. This produces a Catch 22 situation similar to the single driver's frequency response problem. To reduce the demands on the tweeter or mid-range units and to prevent multiple source effects by overlapping frequency response near crossover, the dividing network should have a sharp cut-off, *but* a sharp cut-off produces more phase shift and impairs transient response at crossover frequencies. The designer can minimize electrical phase shifts by choosing a simple crossover network, or he can eliminate them entirely with a single cone speaker.

A law of manufacturing that favors the single cone speaker says, "What you don't put on doesn't cause you any trouble." Tweeters and mid-range speakers are more susceptible to damage from sharp transients than full range speakers are, and crossover capacitors can change value or fail with age. Any reactor in the speaker circuit affects transient response. These facts suggest that the designer of a speaker system should weigh the advantages of any extra component against the possible problems it can cause before wiring it into the circuit.

A single speaker offers two advantages that even the most tone-deaf designer can appreciate: the simplest, quickest hook-up and the lowest cost. Several years ago E. J. Jordan, an English audio engineer, made an analysis of two popular English speaker systems and showed that the 3-way system he examined provided an extra half octave of bass and slightly better high frequency response than an 8 in. full range speaker in a reflex box—but at six times the price. Since he made that examination, several forces have contributed to the popularity of multiple driver systems despite their higher cost.

Distortion is one of the most commonly cited reasons for the simple speaker's loss of favor. When a single cone is expected to reproduce the entire frequency range, there is a good chance that the large cone movements necessary to produce the bass will modulate the treble. One way speakers do this, some engineers say, is by the Doppler effect.

You hear the Doppler effect when an ambulance approaches you with its siren blaring: you notice a rise in pitch until the ambulance passes, and then the pitch gets lower. Theoretically, as a speaker cone moves forward to pump out a bass note, it can produce the Doppler effect by crowding the higher frequency waves together as the ambulance does on its approach. When the cone moves to the rear, it can reduce the wave frequency by moving the source away from the listener. One obvious way to eliminate this or any other kind of bass/treble-produced distortion is to divide the frequency band between two or more drivers, each driver being designed to cover a specific range.

The popularity of compact speaker systems has reduced the appeal of a single cone full range speaker because of the particular demands that a small box puts on the woofer. To yield satisfactory bass response in a small enclosure, the closed box woofers have greater mass and longer voice coils than full range speakers have. The mass helps to cut down the high frequency response, and what is left is killed by the inductance of the long voice coil. Such a speaker is good only as a woofer, so most compact box speakers need at least an added tweeter.

As speaker designers and listeners have gained an increased appreciation for good dispersion, they have gone more to small diameter drivers for mid-range and treble. In earlier times, some people bought a wide range speaker, intending to improve it later by adding a separate woofer and tweeter and keeping the original speaker for the mid-range. This often did not work out as well as expected. One reason was that the full range speaker was too large for ideal mid-range performance. Such haphazard arrangements contributed to the increase in sales of factory-made speakers, even though some lower-priced models also suffered from careless design. Drivers for a multiple speaker system should be compatible.

WOOFER-TWEETER COMPATIBILITY

The same principles that affect woofer-tweeter compatibility apply to woofer-mid-range and mid-range-tweeter combinations (Fig. 2-1).

The most important requirement for a woofer and tweeter to work together is that their frequency ranges overlap. How much they should overlap depends on the cut-off rate of the crossover network. As a rough guide, each driver's response should extend smoothly an octave beyond the crossover point. Tweeters, especially, should not be operated down to their cut-off frequency. If overloaded below their normal band, they can be damaged, or they will produce more distortion. Moreover, the network will reduce damping at resonance if the tweeter is permitted to work in that range.

The efficiencies of woofer and tweeter should be similar. If one driver is much higher in efficiency, it can be padded down with a

Fig. 2-1. A line-up of typical high compliance woofers, mid-range drivers, and tweeters.

varible resistance such as an L-pad, but at the cost of some waste of power. If the woofer is the more efficient member of the duo, the problem is more serious because any resistance in its circuit can affect the amplifier's ability to damp the cone at resonance. A resistance in a tweeter circuit causes no such problems because the crossover network will cut the power to the tweeter above its resonance.

In some dead sounding speaker systems where an efficient woofer is coupled to an inefficient tweeter, the mismatch can be traced to electrical incompatibility. Speakers from different manufacturers with rated impedances of 8Ω can have impedances that range from 5 to 10Ω. If the tweeter has a much higher impedance than the woofer, it will draw less current and may sound weak. Sometimes two such tweeters can be connected in parallel to reduce the net impedance of the tweeter circuit, but then the final impedance should be checked to protect the amplifier (Test 2, Chapter 3). It is a good idea to maintain a minimum impedance of 4Ω in order to limit current in the output circuit. For maximum efficiency the woofer and tweeter impedances should be about equal at the crossover frequency.

The most subtle form of incompatibility between woofer and tweeter is in tone color. Some listeners claim that they can detect differences in tone color if the woofer has a paper cone and the tweeter a mylar diaphragm, or if a horn tweeter is used. The suspicion of horn tweeters may be partly based on a visual difference because of the metal horn structure. Normally this structure is well damped so that it does not ring. Anyone prejudiced against "tin" horns should be aware that many woofer frames also add coloration because the frame is not rigid enough or well enough damped. One

Fig. 2-2. These mid-range speakers show some of the wide variety available to speaker builders.

way of testing a woofer for a ringing frame is to thump the cone with your fingernail while you hold the speaker near your ear. You will probably hear two sounds: a relatively dead sound from the cone, and a metallic sound from the frame. Expensive woofers have die-cast frames which produce little or no ringing. All woofer frames are damped to some extent after the woofer is bolted to the speaker board.

One easy way to avoid the problems of woofer-tweeter compatibility is to choose a set of drivers made by the same manufacturer. This is usually safe, though not necessarily the best course to follow. It *is* better than buying odd woofers and tweeters and wiring them together in hopes of finding a good combination.

HOW TO CHOOSE A MID-RANGE DRIVER

In many 3-way speaker systems built by amateurs, the mid-range driver is the neglected member of the trio. "After all," their thinking goes, "any speaker can reproduce mid-range!" The truth is that although almost any speaker can produce some kind of sound over the mid frequency range, this is the wrong place to compromise. It is in the mid-range that the human ear is most sensitive to peaks, dips, and distortions. It is better to have limited low and high frequency response with a smooth mid-range than to have excellent highs and lows with a peaky mid-range (Fig. 2-2).

If amateurs often place too little importance on good mid-range sound, so do some manufacturers. Cheap mid-range speakers appear to have had no engineering time spent on them except for the planning needed to stamp out a solid metal frame to enclose the back. An enclosed back is a good thing for a mid-range speaker since it eliminates the necessity of building a subenclosure behind it, but the air volume *behind* the cone should be calculated to produce a frequency of resonance that is satisfactorily low at a Q of about 1 or less. Moreover, the designer must control the reflections in the space behind the cone so that they do not produce an uneven response.

The table in Fig. 2-3 shows the characteristics of six mid-range drivers. These six show the range of variation found in testing a dozen mid-range speakers. A look at this table will show you that different manufacturers do approach the design of a mid-range speaker in different ways. Some simply appear to ignore the problems.

Some of the differences are obvious. First, the size of the speakers varies from a small dome radiator of about 1½ in. in diameter to an 8 in. cone. Secondly, resonance occurs at various frequencies from slightly below 200 Hz up to 600 Hz.

Going by the rule of thumb given for woofers-tweeters, the fourth column in Fig. 2-3 shows the minimum crossover frequency at 2f, twice the frequency of resonance. The next column gives the high frequency limit of the driver if its use is restricted to the range where the cone is smaller in diameter than the wavelength of sound. Look at the data for speaker #6, the 8 in. mid-range speaker. By the design methods outlined here, we find that it has a usable range of 1200 to 2000 Hz, while the first speaker listed has a range of 800 to 9000 Hz. As you can see, a single test for resonance and a quick check with a tape measure can show quite a bit about the desirability of a mid-range speaker.

But the difference is much greater than even the usable frequency range figures suggest. Look at the last column, the character of the frequency response curve near resonance. While the 1½ in. driver has a smooth response at a lower frequency, the 8 in. driver with its small magnet and high Q shows a 12 dB peak. So if you were using a 6 dB per octave crossover network, you would have to raise the crossover frequency to eliminate the peak, probably to about 2400 Hz. But if we want ideal dispersion, the speaker is only good to 2000 Hz! What's the use?

At this point someone is sure to say, "This is unfair criticism of the low-priced mid-range speaker." After all, a single 8 in. cone is expected to perform far above 2000 Hz, so why limit a mid-range unit to that range? The answer is that if you do go to the extra expense and complexity of a 3-way system, you should be more demanding on the performance of the drivers or you may end up with a system that is in no way better, and some ways worse, than a single speaker. Even a cheap full range speaker will have a fundamental resonance that occurs at a much lower frequency than any cheap mid-range driver has. If the full range speaker is installed in a large enough subenclosure with the right amoung of damping mate-

SAMPLE	PISTON DIAMETER (IN.)	RESONANCE (Hz)	MINIMUM CROSSOVER FREQ.—2f (Hz)	APPROX. FREQ. WITH WAVELENGTH EQUAL TO DIAMETER	COMMENTS	FREQ. RESPONSE AT RESONANCE
1	1½	400	800	9000	DOME WITH SMALL VOLUME BEHIND IT.	SMOOTH.
2	3	250	500	4500	Q APPROX. 1; DEEP SHELL ON BACK.	VERY SMOOTH.
3	3¼	200	400	4000	DEEP SHELL: Q APPROX. 0.5	ROLL-OFF STARTS ABOUT 2 OCTAVES ABOVE F.
4	3½	300	600.	4000	DEEP SHELL ON BACK.	4 dB PEAK.
5	3½	600	1200	4000	APPEARS IDENTICAL TO #4—BUT HAS SEALED METAL FRAME WITH MUCH LOWER CUBIC VOLUME BEHIND CONE: HIGH Q.	A PEAK OF AT LEAST 12 dB AT RESONANCE.
6	6½	600	1200	2000	SMALL MAGNET: LOOKS LIKE CHEAP 8 IN. SPEAKER WITH SEALED FRAME.	12 dB PEAK AT RESONANCE.

Fig. 2-3. Characteristics of various mid-range drivers.

rial, it will outperform the cheap mid-range driver. There is no point in going to a 3-way system unless you can improve the mid-range sound as well as the high and low frequency performance of your system. Get a decent mid-range driver or none.

If you study the table of data for the six mid-range drivers, you will see subtle differences in the way the better drivers are designed to cover their frequency range. For example, compare the characteristics of #2 with those of #3. Their resonances are similar in frequency, but quite different in response shape because of their difference in Q. The response of #2 holds up well right down to the point of resonance, then rolls off gradually. Because this speaker has almost no peak in response at resonance, it can be used with a network that crosses over down to its 250 Hz resonance without peaky response problems, but its power rating might be reduced by that arrangement. The next speaker, #3, has a lower Q (0.5) that causes it to roll off in response, starting at a point far above the frequency of resonance. The designer used a heavy magnet to damp the cone, probably thinking that the roll-off would permit a less complex and expensive crossover network. Either driver should work well in a speaker system designed to match its characteristics.

As mentioned earlier, the small full range speaker or woofer is a possible source of mid-range drivers other than the models offered for that purpose. Many high compliance 4 in. or 5 in. speakers have a fairly smooth response and will perform well in the mid-range. An advantage of these speakers is that you can build a subenclosure large enough to keep the resonant peak so low in frequency that the crossover network can omit it. You can also experiment with damping materials to get the minimum interference or peaking from internal reflections. If you have testing equipment, you should test the speaker by the methods outlined in Chapter 3 to arrive at an enclosure volume that is right for the crossover frequency you will use. If you don't have testing equipment, use a 4 in. speaker and put a generous volume of air behind it, generous at least in comparison to the typical mid-range speaker. One excellent way of treating such mid-range drivers is to put a tube behind the driver with a cone-shaped reflector in the end to break up the backwave from the cone, and then to fill the tube with damping material.

And here is a final concession to anyone who wants a 3-way system and can't, or won't, buy a high quality mid-range speaker. You can use any cheap open-back speaker and build a subenclosure for it. You will get better sound that way than by buying a cheap, cramped-back mid-range speaker.

HOW TO CHOOSE CROSSOVER FREQUENCIES TO MATCH DRIVERS

Even if you cannot measure frequency response, you can follow some rough rules of thumb in setting crossover frequencies. If the speaker you are using for the low frequency range was designed to be a full range speaker, you can probably limit your additional drivers to a tweeter, and the crossover point will not be particularly critical. For woofers, the following table shows how the rated diameter of the speaker is a good indication of where you should cross over to a smaller speaker:

Woofer Size	Crossover Frequency
8 in.	2000 Hz
10 in.	1500 Hz
12 in.	1200 Hz
15 in.	1000 Hz

For compact box woofers with heavy cones and roll suspensions, these are upper limits; the best choice may be well below the figure listed. If you have no facilities for measuring frequency response, you can check the impedance curve. If possible, try to set the crossover frequency below the point where the curve rises significantly; however, this may not be practical with low resonance woofers.

Try to set the crossover frequency for the tweeter so that the tweeter's resonance or cut-off will lie an octave below the crossover. This puts considerable demand on the kind of tweeter you can use in a 2-way system, so large woofer/small tweeter combinations are rare. If the crossover point is set at 1000 Hz, the tweeter should have a resonance of about 500 Hz, which would be a rare tweeter. More often the recommended crossover point for tweeters is between 2000 and 5000 Hz. This recommendation, coupled with the kind of large woofers often used, explains why we have so many 3-way systems.

To set the crossover points in a 3-way system, follow the same kind of guide lines outlined in Fig. 3-4, with proper consideration of the characteristics of your woofer and tweeter.

CROSSOVER NETWORKS

The ideal crossover network would divide the signal into frequency bands and deliver it to the speakers to produce a uniform output at all frequencies. This ideal system would produce no phase distortion and have no effect on the amplifier. Like the perfect speaker, this ideal network does not exist.

One common misconception about crossover networks is that they chop off the signal to a driver at a certain frequency. Even the sharpest cut-off networks do not do that. Practical crossovers feed signals to drivers far beyond the desired cut-off point.

The simplest kind of frequency divider is a high-pass filter put in series with a tweeter to block the bass, as shown in Fig. 2-4. A

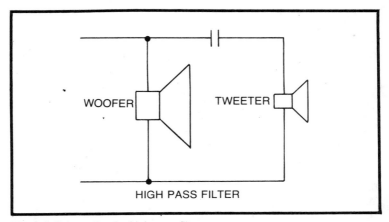

Fig. 2-4. Schematic of a high pass filter.

capacitor has more reactance to low frequencies than to high frequencies, so a single capacitor can act as a high pass filter. One rule of thumb is to choose a value that will offer a reactance to the signal at the desired crossover frequency which is equal to the impedance of the speaker. Such a filter reduces the voltage across the speaker voice coil to 0.707 and the power to 0.5 that of the unfiltered circuit at the same frequency. The response should then be down 3 dB. For any crossover network, the crossover frequency is usually defined as the point where the response is down 3 dB.

A single capacitor high-pass filter offers the least possibility of network problems for the amplifier, but it should be low enough in capacitance to put the crossover point well above the tweeter's resonance. With the capacitor in the circuit, the amplifier's direct coupling to the tweeter is broken and damping at resonance is impaired.

The wire coil has a frequency characteristic, *inductance*, that is opposite to that of a capacitor. When an alternating current is applied to a coil, that coil sets up an electromagnetic field which opposes any change in the current. The coil, called an *inductor* or *choke*, has more reactance to an alternating current than to a direct current. This reactance varies directly with the frequency of the current: the more rapid the change of direction, the greater the reactance. An inductor chokes the highs, but passes lows. To get the right value of inductor for a low-pass filter, use the same rule as for capacitors: choose a choke which offers the same reactance at the crossover frequency as the impedance of the speaker at that frequency.

When a choke and a capacitor are put in series, a band of frequencies must be passed if the crossover points are to be spaced to permit the desired pass-band. This kind of filter can be used to feed the mid-range unit in a 3-way system, as in Fig. 2-17.

Crossover networks can be classified by the kind of speaker wiring circuit (series or parallel), and by the sharpness of the cut-off rate to the various drivers. The naming of series and parallel circuits often causes confusion. When the crossover components are in series with the speakers, the circuit is called a parallel circuit, and vice versa. The sensible key to the names is that in a parallel circuit, the speakers are in parallel with each other; in a series circuit, the speakers are in series and an inductor is in series with a capacitor across the line.

Figure 2-5 shows examples of series and parallel circuits. The parallel circuit is generally recommended over the series circuit. With series circuit, there will be a frequency at which the capacitance reactance will have an equal but opposite phase angle to the inductive reactance. At this point, the reactances will cancel and the amplifier will "see" a direct short, making a hole in the response curve.

The cut-off rate depends on the number of components in each leg of the network, and varies with frequency. Its slope is more gradual near the crossover point than beyond that frequency. For example, a network that includes one element per frequency band is called a 6 dB per octave network, after the ultimate slope beyond crossover. At crossover, the slope is only 3 dB per octave. A network that includes two filter elements (a capacitor and an inductor) per section is called a 12 dB per octave network and has a 6 dB slope at crossover. And so on.

To make a 12 dB network, you can combine a series high-pass or low-pass filter (as in a simple network) with a shunt element. In the tweeter circuit, you would wire a capacitor in series with the tweeter and a choke across the tweeter. The capacitor blocks the highs, and the choke provides a low impedance path around the tweeter for the lows. This double action filter has the desirable effect of increasing the rate of cut-off, but it also causes a complication by increasing the rate of phase change over the frequency band. At crossover, there is a 180 degree phase shift which requires that drivers in adjacent frequency bands be connected out of phase. For example, a mid-range driver would be connected out of phase with both woofer and tweeter.

The debate about what kind of crossover network is best will probably continue as long as passive networks are in use. Some designers prefer to add no inductive loads to the output circuit except that of the speaker. Such additional elements will have some effect on woofer damping, so extremists on this point choose the simple single capacitor high-pass filter. Its simplicity and low cost are particularly appealing for inexpensive speaker systems.

An advantage of a more complete network is that it eliminates high frequency signals to the woofer and thus conserves power. Another reason for cutting off highs to the woofer is to depress the spike that many woofers have in their response curve before cut-off. A bad peak may influence a designer's choice of network cut-off rate.

Considering all the arguments, the best recommendation for the amateur builder of crossover networks seems to be the 6 dB per octave network. This kind of network is simple and cheap, has a little more than a third of the inductance of a more rapid cut-off filter, and has a more gradual phase shift at crossover and an ultimate phase difference of 90 degrees between circuit branches. The only disadvantage to this type of network is that its gentle slope permits both drivers to operate further beyond the crossover point. To prevent tweeter overload, a compromise network can be devised by adding a shunt choke across the tweeter branch.

If the crossover point is to be at 500 Hz or below, it may be more practical to use a commercial iron-core inductor than a homemade air-core coil. Note that "air-core" refers to any choke having a non-magnetic core, such as a wooden dowel rod. Inductors with iron cores are considered to cause more distortion than air-core chokes, but commercial iron-core chokes do work satisfactorily. An air-core choke large enough to work below 500 Hz would require a lot of wire.

HOW TO USE PIEZOELECTRIC TWEETERS

These tweeters are usually more efficient than other drivers, having such a high impedance at low frequencies that they are effectively removed from the amplifier circuit. If the tweeter output level matches that of the other speakers, the piezo-electric tweeter can be wired right to the amplifier output line at the input to the crossover network. This kind of hook-up applies full damping to the tweeter. Although this direct connection gives the best transient response, the tweeter may need a level control.

Figure 2-6 shows how to add various series and shunt elements to control both the output level and the frequency response of one of these "crystal" tweeters. The simple series resistor can be replaced by a variable resistance, perhaps a 500 to 1000Ω potentiometer, but with some loss of high-end response. A voltage divider such as the one shown in Fig. 2-7 works better. In one case, when the amplifier has only marginal stability at ultrasonic frequencies, a series resistance of about 20 to 30Ω is beneficial. The piezo-electric tweeters have a low impedance at 50 to 100 KHz, which can upset some amplifiers. A low value resistance in series with the tweeter will have little effect on frequency response, but it will stabilize the amplifier.

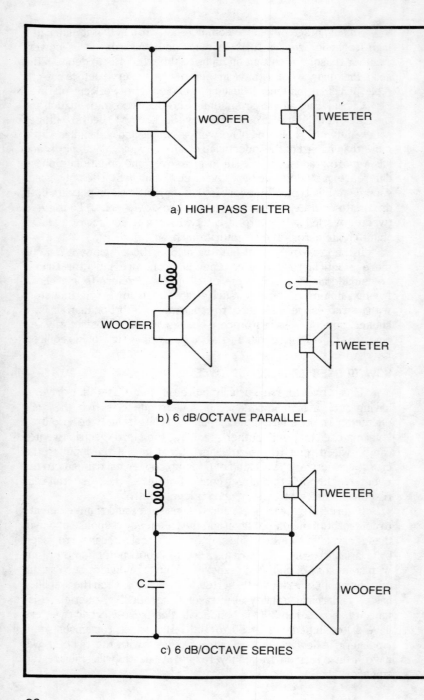

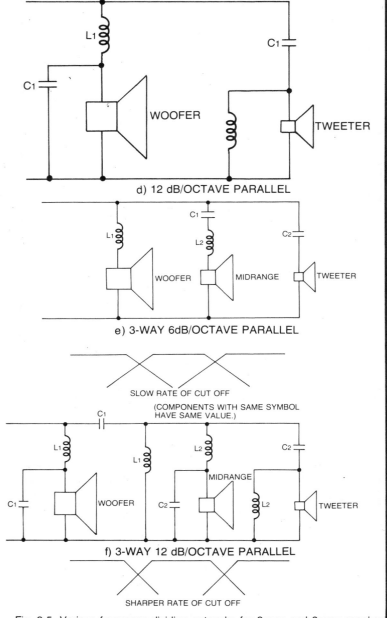

Fig. 2-5. Various frequency dividing networks for 2-way and 3-way speaker systems.

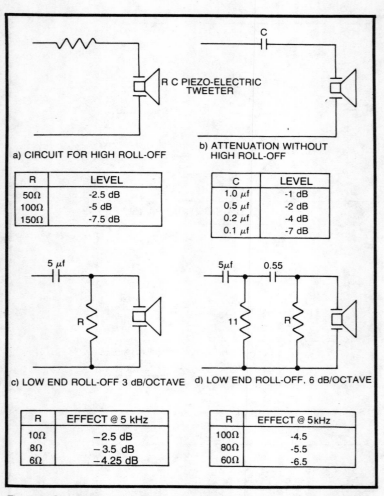

Fig. 2-6. Circuits for piezo-electric (crystal) tweeters. Performance with filter components may vary somewhat, depending on system. Tables show how various series and shunt elements affect the output level and response of a piezo-electric tweeter. (Data from Piezo Ceramic Products Group, Motorola, Inc.)

You can combine the filter circuits in Fig. 2-6 with a level control to get the response and tonal balance you want. The roll-off figures shown with the circuits are typical values; the actual performance may vary according to your total circuit. Although these tweeters will work without a crossover, some users report cleaner sound when a sharp low end cut-off filter is used, such as the filter shown in Fig. 2-8.

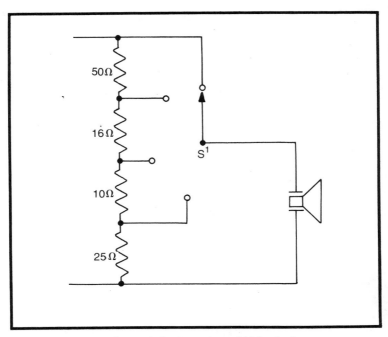

Fig. 2-7. Schematic of control circuit courtesy of Motorola, Inc.

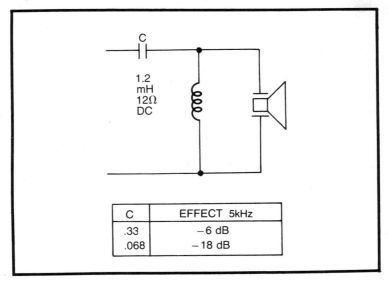

Fig. 2-8. Piezo-electric circuit for low end roll-off.

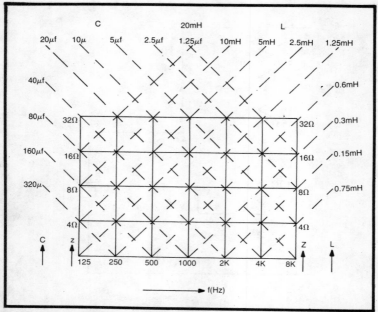

Fig. 2-9. Crossover network design chart. Approximate values of inductance and capacitance for speakers with ratings from 4 to 32Ω.

HOW TO MAKE A CROSSOVER NETWORK

Figure 2-9 shows the correct values of filter components to use with speakers of various impedances for any crossover frequency from 125 to 8000 Hz. To use the chart, follow the vertical line up from the desired crossover frequency to the horizontal line that represents the speaker impedance. Ideally, this impedance value should be the driver's impedance in the crossover region rather than just the rated impedance. For best efficiency, the speakers should have equal impedance.

Here's an example: a woofer and a tweeter each have an impedance of 8Ω. The woofer's response is good up to about 2000 Hz and the tweeter's down to 400 or 500 Hz. A logical point for crossover is 1000 Hz, so we follow the middle vertical line up to the second horizontal line (8 Ω). Then follow the dashed line to the upper left to get the value of capacitance, which is 20 μf (1 μf is 1 microfarad = 0.000001 farad), and the dashed line to upper right to get the inductance, which is 1.25 mH (1 mH is 1 milliHenry = 0.001 Henry). The values given here for inductance are rounded off from calculated values, but the accuracy of the figures is more than adequate.

The best kinds of capacitors for crossover networks are the mylar, paper, or oil-filled capacitors, but the more common non-polarized electrolytics can also be used. Polarized electrolytics are designed for dc circuits, so they are not appropriate for audio circuits. The preferred kinds are very expensive and practical only where you need values of a few microfarads, unless you can find them on the surplus market.

If you cannot find a capacitor of the exact value needed, you can wire several smaller capacitors in parallel. Remember that when capacitors are connected in parallel, the net capacitance of the group is equal to the sum of the individual values.

If you can find some magnet wire, you can make your own air-core chokes (Fig. 2-10). Electronics stores have small-diameter

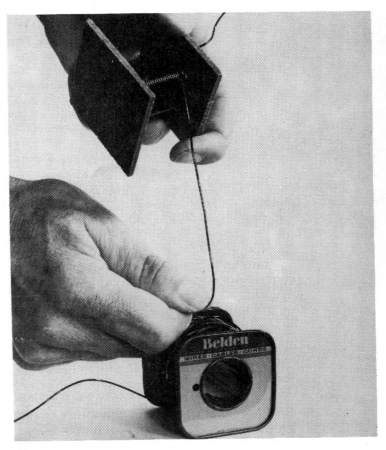

Fig. 2-10. Winding a choke by hand.

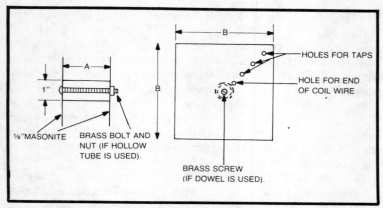

Fig. 2-11. Forms for homemade coils. See Fig. 2-13 for dimensions.

wire, such as #22 gauge, which can be used in single strands for small chokes or with triple strands connected at their ends for larger chokes. The most useful wire size is #18 gauge, one strand of which is equal to about three of #22 gauge, because its diameter is great enough to keep the resistance of the inductor low. In a choke we want the reactance to be as purely inductive as possible, not resistive. To find #18 gauge wire, check out electric motor repair shops; they often use #16 to #18 gauge wire to rewind motor armatures and will either be able to sell you some wire or to tell you where to get it.

Make up coil forms similar to those shown in Fig. 2-11. The coil forms can be temporary forms held together by brass screws or by a brass bolt and nut, or by being permanently glued together (Fig. 2-12). The temporary forms can be used again and again, but the permanent forms offer protection for the coil if the speaker system is to be moved frequently. A tapped coil has sufficient stability for most purposes.

The coil form center piece can be a 1 in. wood dowel, a section of a plastic pill container, or even a piece of broom stick. Use ⅛ in. Masonite or hard plastic to make the flanges. Avoid iron screws or bolts and install the choke away from heavy iron components. Do not stack coils; mutual inductance can change the effective inductance.

Note that small holes must be drilled into one flange of each coil form to permit wire exits. For a simple coil, only the center hole is necessary, and it should be placed so that the wire will exit against the core piece. If you plan to wire a tapped coil, you can drill the extra holes as needed.

Now to wind the coil. Thread an inch or more of the magnet wire out through the center exit. Then start winding, making sure that the first loop is placed tightly against the flange with the hole in it. Try to make each turn lie flat against the previous turn. Wind two or more layers, then use a strip of masking tape to cover the wire and hold it in place. As you continue winding, try to keep the layers flat. Although scramble-wound coils will work just as well (unless you leave gaps between adjacent turns by too much cross winding), you can predict the final inductance better if you do a neat job.

The normal way to make a coil to specifications is to add more than the calculated number of turns, and then to remove turns until the measured inductance reaches the desired value. A more practical way is to make a tapped coil. When you have completed about 75% of the winding, stop at the end of that layer and bare a short section (about ¼ in.) of magnet wire by scraping it with a sharp knife. Solder a piece of copper wire to the bare place on the magnet wire, and carry the wire out through a prepared exit hole in the flange. It is a good idea to make the taps with insulated hook-up or bell wire, using a different insulation color for each tap so that they can be easily identified later. You can use tape to insulate the soldered connection, but the lacquer on the adjacent magnet wire would be sufficient. Wrap another 10% to 15% of the coil and make another tap. When you are ready to use the coil, connect it in your network at each tap and choose the one that gives the best performance. Or you can measure the inductance of the coil at each tap as described in Test 16, Chapter 3. When testing or wiring the coil, remember that the taps will be closer to one end of the coil than the other, so be careful to identify each of the two coil end wires. Failure to do so could cause such a major mistake in wiring that the inductance might be only about 10% of the value you wanted.

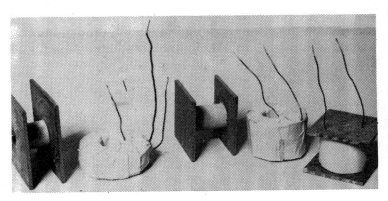

Fig. 2-12. Left to right: A temporary coil form, a tapped coil, a permanent coil form, a coil removed from the form and taped with masking tape, and a coil used in its permanent form.

VALUE OF INDUCTANCE	COIL FORM DIMENSIONS	
UP TO ABOUT 2 mH	A	1 IN.
	B	4 IN.
ABOVE 2 mH	A	1-½ IN.
	B	4 IN.

INDUCTANCE (mH)	APPROX. NO TURNS	APPROX. WT. WIRE
10	625–700	2 LBS.
5	450–500	1 LB.
2	300–350	
1	200–250	½ LB.
0.5	150–175	
0.25	100–125	¼ LB.
0.1	65–90	

Fig. 2-13. Coil winding data for #18 gauge enamel coated magnet wire or a similar larger gauge (#17).

This is how one coil turned out. The desired inductance was 1 mH, and a partial roll of #17 gauge wire weighed about 10 oz. on a postal scale. This wire is slightly larger in diameter than #18 gauge, but not so much as to require another chart than the one in Fig. 2-13 which shows that a half pound of wire should yield 200 to 250 turns and given an inductance of 1 mH. When unrolled and measured, the #17 gauge wire was about 95 ft. long. The coil was started and wound rapidly, with no care taken to make the layers perfect. Two taps were made: the first one after about 70 or 75 ft. of wire had been wound (175 turns), and the second at 190 turns. The total number of turns was 203, subject to the wandering attention of the coil maker and possible counting errors. Here's how it measured from center wire to:

> 1st tap.......... 0.8 mH
> 2nd tap......... 1.0 mH
> End of coil.... 1.15 mH

Notice that even though a larger wire size was used and the winding was somewhat scrambled, the inductance values came out fairly close to the desired one. Air-core coil winding is easy. You can get the values you want if you have test equipment or know someone who will measure it for you. The tests in Chapter 3 use ordinary audio test instruments, but inductance is usually measured with an impedance bridge.

If you don't mind working with junk, the cheapest way of all is to get your crossover components from a salvage yard. These places often have discarded telephone equipment or industrial electronic items that contain good quality paper or oil-filled capacitors, and you can find magnet wire in the electric motors of abandoned appliances.

HOW TO DESIGN A SIMPLE
6 dB/OCTAVE 2-WAY CROSSOVER NETWORK

The basic schematic for a simple 6 dB/octave 2-way crossover network is shown in Fig. 2-14. First, measure the impedance of the speakers. If you do not have all of the equipment necessary for running impedance tests, such as Test 2 in Chapter 3, you can measure the dc resistance of the voice coil and add 1/3 of the reading you get. For example, if you read a dc resistance of 6 Ω, the impedance is 8 Ω. If possible, do make Test 2 at the crossover frequency and use the value you get for the impedance.

Next, go to the design chart in Fig. 2-9 and find the values of the components. If you have 8 Ω speakers and you want to set the crossover frequency at 2000 Hz, you will follow the 2000 Hz line up to the horizontal line for 8 Ω. The diagonals that meet there show a choke of 0.6 mH for the woofer and a 10 μf capacitor for the tweeter.

If the crossover frequency falls between the lines shown on the chart, you can calculate the right values from these formulas:

$$L = \frac{X_L}{2\pi f} \quad \text{and} \quad C = \frac{1}{2\pi f \, X_C}$$

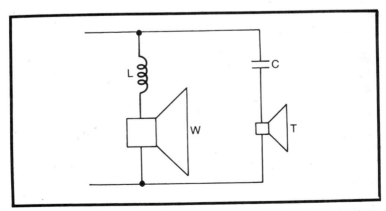

Fig. 2-14. Simple 6 dB/octave 2-way crossover network. Use values from chart.

Where:

    L = value of inductance
    X_L = inductive reactance at crossover frequency = speaker Z
    f = crossover frequency
    C = value of capacitance
    X_C = capacitive reactance @ crossover frequency = speaker Z

HOW TO ALLOW FOR VOICE COIL INDUCTANCE

Because of their long voice coils, modern high compliance, long throw woofers have considerable self inductance which, in effect, is part of the crossover network whether you want it or not. One advantage of a parallel network having an inductor in series with the woofer is that you can make allowances for the voice coil inductance when you calculate the value of the choke you need.

After you have checked the impedance and calculated the inductance for the choke, apply Test 15 to the woofer to measure its voice coil inductance. Then subtract the value of the voice coil inductance from the inductance needed in the crossover circuit. If the crossover frequency is at the top of the woofer's range, you will sometimes find that the voice coil inductance is equal to the value of the choke, which means you can eliminate the choke from the circuit.

If you do not have the equipment to run Test 15, approximate the right choke value by measuring the dc resistance of the voice coil and using that figure instead of impedance when you calculate the value of the choke to use. This procedure is based on the assumption that, at a typical crossover frequency, the difference between the voice coil dc resistance and impedance is caused by voice voil inductance. This assumption works about as well as the test for voice coil inductance. In tests on four woofers, the percentage difference in choke values obtained by the two methods was 0%, 5.8%, 7.5%, and 8%.

It should be noted that adjustments of choke values to allow for voice coil resistance can be used only with parallel networks. Some experts prefer series networks for various reasons. The projects described in this chapter will include both types of network.

PROJECT 1: A SIMPLE 2-WAY SYSTEM

For this first project (Fig. 2-15) we chose an 8 in. woofer and a 1½ in. tweeter for which we will design a 6 dB/octave 2-way parallel crossover network. By going through the steps, you can apply the same process to any two speakers you want to use.

The speakers are rated at 8 Ω, but we ran impedance curves according to Test 2, Chapter 3. The tweeter has uniform impedance

Fig. 2-15. Project 1. Front view of a simple 2-way speaker system.

throughout the crossover range, but the woofer's curve dips to about 5.5Ω at 200 Hz, stays almost flat to about 400 Hz, then starts rising sharply to 11Ω at 800 Hz, 18Ω at 2000 Hz, and about 60Ω at 10,000 Hz. Theoretically, we should set the crossover point at 400 Hz or below, in the flat part of the woofer's impedance curve, but that is impossible for a 2-way system using a small tweeter. Checking the tweeter, we find an impedance rise just below 1000 Hz, so we will set the crossover frequency at 2000 Hz.

The design chart only gives an approximation of the correct choke value for an 18Ω impedance at 2000 Hz, so we must calcualte it.

$$L = \frac{18}{(6.28)(2000)}$$

$$= 0.00143 \text{ H or } 1.43 \text{ mH}.$$

This value would be exact if it weren't for the voice coil inductance. A check of the woofer's inductance by Test 15 shows it to be 1 mH. We should subtract 1 mH from 1.43 mH to get the right choke value, or 0.43 mH.

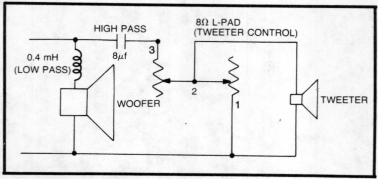

Fig. 2-16. Project 1. Schematic.

As a rough test of the calculations, we can apply the substitute method of getting the choke value by measuring the voice coil dc resistance. This turns out to be 5.2Ω, so the choke should be:

$$L = \frac{5.2}{(6.28)(2000)}$$
$$= 0.00041 \text{ H or } 0.41 \text{ mH.}$$

You can make this choke by winding about 130 turns of #18 gauge magnet wire on the smaller (1 in. by 1 in.) coil form shown in Fig. 2-11.

The tweeter's impedance being 8 Ω at 2000 Hz, the value of the tweeter capacitor can be obtained directly from the chart and turns out to be 10 μf. However, tests indicated that 8 μf were adequate and it is permissible to use a small value capacitance if it does not produce a hole in the response curve. A greater value should not be used because one purpose of the capacitor is to protect the tweeter. The schematic diagram of the system is shown in Fig. 2-16. The construction details for this project appear in Chapter 15.

Note: Tests on this woofer's "big brother" (10E8540 CTS, the 10 in. model) indicate that it could be substituted for the woofer used here without any change being made in crossover network components, althout it would require a larger box with an internal volume of about 1.5 cu. ft.

HOW TO DESIGN A 6db/OCTAVE 3-WAY PARALLEL NETWORK

This network is diagrammed in Fig. 2-17.

First, choose the crossover frequencies. For convenience in using the chart for this example, we will choose 500 Hz and 4000 Hz and assume that we have 8Ω speakers. For L_1 we go to the chart and follow the 500 Hz line up to the 8Ω horizontal line. The diagonal line to the right tells us that the choke value should be 2.5 mH. If we knew the voice coil inductance of the woofer, we would subtract its value from 2.5 mH. For example, if the woofer had a voice coil inductance of 1 mH, we would use a 1.5 mH choke.

Because of interaction between the elements in the mid-range filter, the effective crossover frequencies for it are not exactly those perdicted by formula. To compensate for the shift, you should subtract the lower frequency from the upper one to get the frequency for the coil.

Since we have a 500/4000 Hz network, we would subtract 500 from 4000 to get 3500 Hz, the figure to use in calculating the coil value. This formula is:

$$L = \frac{Z}{2\pi f}$$

So, in this case:

$$L = \frac{8}{(6.28)(3500)}$$

$$= 0.00036 \text{ H or } 0.36 \text{ mH}.$$

Notice that if we had used the chart for this value, it would have been 0.3 mH. Actually, the difference may not be significant; voice coil inductance can easily be greater than the error.

For the mid-range capacitor, we want to find the right value to cut off the lows at 500 Hz. Although the chart gives us 40 μf, we know that there will be some shift downward in the effective crossover frequency and that we can compensate for this shift by calculating the theoretically correct value. To get the correct frequency

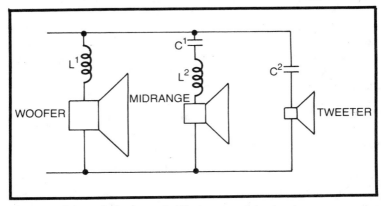

Fig. 2-17. Schematic for 3-way 6 dB/octave parallel network.

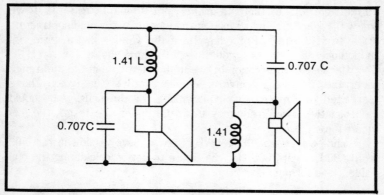

Fig. 2-18. Schematic for 12 dB/octave network. Multiply values from chart in Fig. 2-9 by factor shown here.

figure to use in calculating the mid-range capacitance, use the formulas:

$$f' = \frac{1}{\frac{1}{f_L} - \frac{1}{f_H}} \quad \text{and} \quad = \frac{1}{\frac{1}{500} - \frac{1}{4000}} = 571 \text{ Hz}$$

$$C = \frac{1}{2\pi Z \, 571} = \frac{1}{(6.28)(8)(571)} = 0.0000348 \text{ f or } 35 \, \mu\text{f}.$$

Finally, we find the value for the high pass filter to the tweeter, C_2, by locating the 4000 Hz line on the chart and following it to the intersection of the 8 Ω horizontal and the 5 μf diagonal.

To summarize the network problem posed above for an 8Ω network with crossover points at 500 and 4000 Hz, we would need the following components: $L_1 = 2.5$ mH; $C_1 = 35$ μf; $L_2 = 0.36$ mH; and $C_2 = 5$ μf. Even though some of the values were calculated, the chart tells us quickly whether we are in the right ballpark.

HOW TO DESIGN 12 dB/OCTAVE CROSSOVER NETWORKS

These networks require much more magnet wire and have been criticized for their greater phase shift at crossover, but they do offer a sharper cut-off slope. If you want to try your hand at a 12 dB network such as the one in Fig. 2-18, here is the design procedure.

Choose the crossover frequency; 1000 Hz will be used for this example. Find the component values for a 6 dB/octave network from the chart in Fig. 2-9. For 8 Ω speakers, these would be 1.25 mH and 20 μf.

Multiply the inductor value by 1.41, which makes it 1.76 mH; and the capacitor value by 0.707, which makes it 14 μf.

Make up two coils, each with an inductance of 1.76 mH; and purchase two capacitors, each with a value of 14 μf. Install these components in the network diagrammed in Fig. 2-19.

HOW TO DESIGN A 3-WAY 12 dB/OCTAVE CROSSOVER NETWORK

Choose the crossover frequencies to match your drivers. Here we will choose 500 Hz and 4000 Hz for convenience in using the chart in Fig. 2-9.

Look at the diagram in Fig. 2-20. Notice that if a line were drawn down through the middle of that diagram, just to the right of the second inductor, we would have two circuits, each with the same number of components as a 2-way crossover network. To simplify the design, you can consider yourself to be designing two circuits by the same procedure as that shown above for the 2-way 12 dB/octave network, and then fitting them together to make a 3-way crossover network.

First let us consider the network that separates the woofer from the other drivers. To find the right values, we go to the chart and see that the 6 dB/octave figures for 500 Hz would be 2.5 mH and 40 μf. Multiplying these by the appropriate values, they become 3.5 mH and 28 μf. So we draw in two 3.5 mH chokes and two 28 μf capacitors.

Next for consideration is the 4000 Hz crossover. Notice that for the mid-range driver, the low frequency cut-off has already been arranged by the pair of 3.5 mH choke and 28 μf capacitor combinations. The values needed now for the mid-range and tweet-

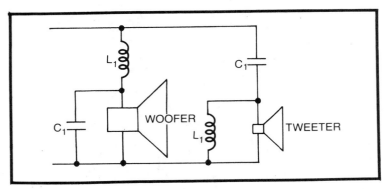

Fig. 2-19. Schematic for 12 dB/octave parallel network.

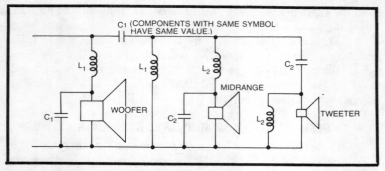

Fig. 2-20. Schematic for 3-way 12 dB/octave parallel network.

er will be identical, but they will go into the circuit in reversed position so that the mid-range will cut off at 4000 Hz and the tweeter will come on at that frequency. Of course, both will continue to produce sound beyond the cut-off, but at a diminished volume.

Going to the chart, the 4000 Hz components for a 6 dB network would be 0.3 mH and 5 µf. After multiplying these by the appro-

Fig. 2-21. Project 2. Front view of a 3-way speaker system.

priate factors, they become 0.42 mH and 3.5 µf; therefore, for this network, we need two 0.42 mH chokes and two 3.5 µf capacitors.

To simplify calculations, we are assuming that all the speakers had an 8Ω impedance at the crossover frequency. The impedance of each should be measured and the components in its branch of the circuit should be adjusted to match the speaker's true impedance.

PROJECT 2: A 3-WAY SPEAKER SYSTEM

To illustrate how to design a series network, this project has a series crossover filter (Fig. 2-21). This filter has the same number of elements in it that a parallel 3-way network would have, but it gives some extra tweeter protection by the double action of two inductors, the first being across the mid-range driver and tweeter, and the second being across the tweeter. The schematic of the proposed crossover network is shown in Fig. 2-22.

Since we cannot subtract for woofer inductance in this kind of network, we should use a woofer that has little or no rise in impedance at the crossover frequency. In this case, three drivers with 8Ω impedance have been chosen: a 10 in. woofer with a foam suspension, a Peerless mid-range driver, and a Peerless tweeter. A check on the woofer's impedance shows that it is 9Ω at 500 Hz, the desired crossover frequency, but this change from its rated impedance is significant.

The choice of woofer was made on the basis of its low Q as well as of its impedance. This woofer has a Q of about 0.4, so it can be used in a ported box. The box volume should be calculated to give the woofer's flattest, most extended bass response. These box

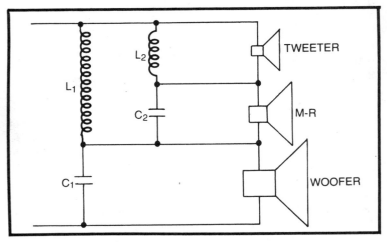

Fig. 2-22. Project 2. Schematic.

calculations will be explained in similar box design problems in Chapter 7.

HOW TO CHOOSE COMPONENT VALUES

The Peerless mid-range driver is the second one listed in the Fig. 2-3. We will set the crossover frequencies at 500 Hz, the lowest permissible ideal crossover frequency, and at 5000 Hz, just slightly above the upper limit. Although the 4500 Hz upper limit is based on cone diameter, the practical upper limit for good dispersion may be much higher. Anyway, 5000 Hz is close enough to the ideal range.

We should first obtain the values of L_1 and C_1 for 500 Hz, because these two components set the dividing point between the woofer and the other drivers. Going to the chart and following the 500 Hz line up to the 8Ω horizontal, we find that L_1 should be about 2.5 mH and C_1 should be 40 μf.

The values for L_2 and C_2 could be found by interpolation from the chart, but that will not be necessary in this case since the frequency is 10 times that of the lower crossover point. For speakers of the same impedance, 8Ω, we can simply divide the values obtained for 50 Hz by 10 to get the right values for 5000 Hz. So L_2 should be about 0.25 mH and C_2 should be 4 μf.

The L-pads which control the tweeter and mid-range driver will be discussed later. Construction details on this crossover design project will be given in Chapter 15.

FINAL SUGGESTIONS

In this discussion of crossover networks, the assumption has been that anyone making a crossover network will test the speakers and use the measured impedance at the crossover frequency in order to match components, but, for anyone without test equipment, a more practical appraoch is simply to use the rated impedance of the speakers. If the speakers are close to their rated impedance at 200, 300, or 400 Hz, the chances are you will get good results. Even though the impedance rises at higher frequencies and requires a larger choke for the low-pass filters, when voice coil inductance is considered; you can go right back to a smaller choke. Of course, not all speakers are correctly rated, so its a good idea to at least measure the dc resistance. Just add about 25% to 30% to that reading, and you will know the correct reated impedance.

Finally, remember that speakers are acoustical devices as well as electrical. A certain value looking right form electrical measurements is no guarantee that it is the optimum value. After calculating the right electrical values, try other components in a similar range, but greater or lower in value.

3

How to Test Speakers and Speaker System Components

The only way you can know if your speakers are performing properly is to test them yourself. Without tests, you are almost completely in the dark with an "orphan" speaker.

The tests described in this chapter are numbered from 1 to 21 for easy reference throughout the book. Combinations of test equipment and aids to make each test are labeled by letter and shown in Figs. 3-6, 7, 13, 14, 16, 19, 21, 26 and 27. For example, to run an impedance curve (Test 2), you would use test set-up A, the same equipment and hook-up that is used for Test 1 (free air resonance).

If you have no audio test equipment, you may be able to rent it. If you do plan to rent, read the tests carefully and have the items you need ready to be tested before you get the equipment.

USEFUL TEST INSTRUMENTS

In addition to its duty as an audio tester, a VTVM (vacuum tube voltmeter) is useful for checking appliances, car batteries, or any other electrical items. Standard VTVM's usually have a low voltage range of 1.5 V full scale, which is adequate for speaker measurements, especially if you are using an amplifier to drive the speaker. (A straight power amplifier with no controls is best.) Special ac voltmeters can give full scale deflection at lower voltages, 0.3 or 0.1 V, or even lower. These are much better for speaker testing, but they cannot be used for dc or resistance measurements.

An audio generator is a high priority instrument. You can make rough tests with a test record, but test records are inconvenient and

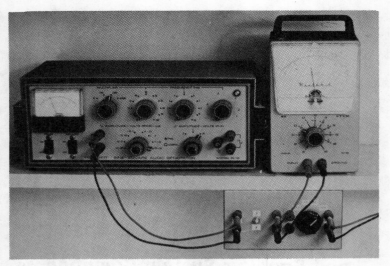

Fig. 3-1. Test instrument coupler, containing connectors and the parts shown in Test Set-up A, is a convenient aid in testing speakers.

their accuracy depends on two mechanical components, both the turntable and the phono cartridge. An audio generator is much more convenient, instantly providing your choice of test frequencies. You can do a great deal with a VTVM and a generator. Almost any audio generator will be adequate for testing crossover networks and designing a closed box speaker system, but for ported box work or for checking distortion, you should have one with good accuracy and stability. The Heath IG-18, shown in Fig. 3-1, is one example of a low distortion generator that is both accurate and stable.

A pocket calculator is a great aid. The scientific models are the most helpful, particularly for computing the low frequency response of a ported box speaker system. Any pocket calculator that has an x^2 and ($\sqrt{x}$) function is extremely useful, particularly if that function can be used to obtain answers for numbers raised to any positive or negative power. A calculator with logarithm keys can solve frequency response problems more quickly than one without, but you can always look up log values in a table.

Some of the tests described here require an oscilloscope. Although an oscilloscope is not necessary to design and build your own speaker system, it does provide additional information that can make the planning more precise in some cases. Most oscilloscopes have a frequency response that is much wider than the audio band, so almost any working oscilloscope will be good enough for speaker tests.

A microphone can be used with a VTVM, or with an oscilloscope to monitor your speaker's output. You may not be able to run really accurate frequency response curves, but you will be able to make rough approximations that will help you to design crossover networks and to match drivers. Choose a high fidelity mike, since some special purpose models have a limited frequency response. Any good brand of dynamic mike will probably have a smoother response than your speakers, but you may need a preamp to get enough output to drive a VTVM. Some inexpensive dynamic mikes are fairly good; condenser mikes are better for speaker testing, but expensive.

Sound-level meters are calibrated directly in decibels, so they do many of the same jobs as a microphone-VTVM combination. You can easily use one to check your speaker's efficiency, frequency range, phasing, and usable sound output power level. A sound-level meter is more portable and convenient, but less versatile than a separate microphone with auxiliary equipment.

This list could be expanded to include distortion-analyzing equipment, tone-burst generators, sweep generators, and so on, but unless you are planning to become a full-time audio technician, the few items mentioned earlier should be adequate.

HOMEMADE TEST EQUIPMENT

You can make some of the equipment yourself. A standard test box, for example, can save calculations, help prevent possible errors, and show some speaker faults. For the cone mass and compliance test, you can make a simple balance with an ordinary ruler and a coin. To test speaker polarity and damping, you can use a battery, switch, and resistor. Instead of trying to find logarithmic graph paper, you can use any ruled paper for impedance curves if you have a slide rule handy.

HOW TO PLOT AN IMPEDANCE CURVE

Materials: Logarithmic graph paper (or ruled paper) and a slide rule.

To make your own log graph paper, you can use a slide rule with a 10 inch scale to make 2-cycle paper for impedance values from 10 to 1000 Hz. If you want 3-cycle paper (10 to 10,000 Hz) or 4-cycle paper (10 to audio frequencies above 10,000 Hz), you can use a slide rule with a 5 inch scale. Note that each slide rule "B" scale has two cycles on it. If you need only two cycles, the paper can be used in the normal vertical position with a 5 inch scale slide rule. For two or more cycles with a 10 inch scale, or three or more with a 5 inch scale, place the long dimension of the paper in the horizontal plane. Any graph paper will have lines ruled in both directions, which simplifies amplitude plotting.

Procedure: Place the B scale across the bottom of the paper and rule in vertical lines at each numbered position on the scale. Number the lines 10 to 100 Hz for the first cycle, 100 to 1000 Hz for the second cycle, 1000 to 10,000 Hz for the third cycle, and so on. Check the value of your highest impedance reading (f_s), and assign numbers along the left vertical margin so that the impedance at f_s will fall somewhere below the top of the page. Label the values at the left margin at 10 Ω intervals: 10, 20, 30, and so on.

Plot the impedance curve by marking points on the graph that show the impedance at the ruled frequency lines. The points on the graph should show a smooth line with a high peak at f_s. If the curve shows any sharp bends, repeat Test 2. Connect the points with a line, but rework the line if necessary to smooth out any minor wiggles.

How to Make a Simple Balance

Materials: Pencil, ruler, and a new, or nearly new, nickel. (A newly minted nickel has a mass of 5 g.)

Procedure: Set the ruler across the pencil so that the pencil is directly under the 6 inch mark. Make sure the ruler is touching a smooth side of the pencil, and not the printing or any other roughness that would make balance tricky. Put the nickel at one end of the ruler and add clay to the other end until they balance. The centers of gravity of the nickel and of the clay should be at equal distances from their respective ends of the ruler. If you need 10 or 20 g., repeat the process and press the 5 g. pieces together to make a single piece.

How to Make a Standard Box

The ideal standard box technique would be to put the speaker inside a box of standard volume, say 1 ft.^{3}, and test it there. The problem is that even a minor air leak would apply great leverage to throw the calculations way off the mark, and a box with a removable panel *would* have air leaks, especially after the panel was replaced several times.

Instead, make an air tight box with no removable panels, but with a speaker hole in one side to match the diameter of your speaker. The speaker is tested on the box, not in it. The box can be used over and over again for any speaker of similar diameter, but you should make a correction for the volume that would have been occupied by the speaker if it had been inside the box.

It is not necessary to make the standard box to any specific volume as long as you know the exact volume for calculations. Some

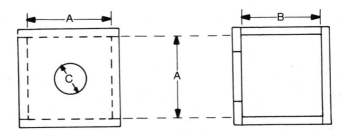

	SPEAKER DATA			STANDARD BOX DATA				
SIZE	TYPICAL PISTON AREA		TYPICAL VOLUME	SUGGESTED DIMENSIONS (IN.)			CUBIC VOLUME (FT^3)	
IN.	IN^2	CM^2	FT^3	A	B	C	ACTUAL	CORRECTED
5	12	77	0.05	8½	3	4	0.125	0.175
4 × 6	12	77	0.05	8½	3	oval	0.125	0.175
6	18	116	0.1	8½	5	5	0.2	0.3
5 × 7	18	116	0.1	8½	5	oval	0.2	0.3
8	28	180	0.15	8½	8½	6¾	0.35	0.5
6 × 9	28	180	0.15	8½	8½	oval	0.35	0.5
10	50	320	0.25	12	9	8¾	0.75	1.0
12	78	500	0.4	13	10¼	10½	1.0	1.4
15	133	850	0.75	18	12¼	13¼	2.3	3.0

Fig. 3-2. Standard test boxes for Test 6.

typical box sizes and dimensions are shown in the table in Fig. 3-2. Choose a box size to match your speaker as shown in this table.

Cut out the parts. Glue and nail the sides together. Glue and nail down the speaker board. The next step is particularly important: caulk the interior joints with a silicone rubber or latex caulking compound (Fig. 3-3). Then glue and nail on the bottom panel. Finally, reach through the speaker opening to caulk the interior joints around the bottom.

How to Make a Polarity/Damper Tester

Materials:

1	SPDT Mini Toggle Switch	Radio Shack #275-326 or eq.
1 pkg.	Banana Plugs	Radio Shack #274-721 or eq.
1	0.33 Ω Resistor	
1 piece	Test Probe Wire, Red	
1 piece	Test Probe Wire, Black	
1 pkg.	Insulated Alligator Clips	Radio Shack #270-378 or eq.
1	Battery Holder	To fit size C-1.5 Volt Battery
1	Small Chassis (Can use tuna can or small pineapple can.)	

Fig. 3-3. Making a standard test box. Caulk all the interior joints before gluing and nailing on the bottom, then caulk the bottom through speaker hole.

Procedure: Using the tuna can as a chassis, fit in the components according to the layout shown in Fig. 3-4. The schematic is given in Fig. 3-19, and the tester is shown in use in Fig. 3-20.

How to Make a Microphone Extension

Materials: Car battery booster clip, piece of aluminum tubing from an old TV antenna, rivets, 3/16 in. bolts, or sheet metal screws.

Procedure: Bend the aluminum tubing into an L shape. Fasten one handle of the battery booster clip to the short leg of the tubing. Flatten the other end of the tubing and drill a ¼ inch hole through it. Tap the hole to fit a camera tripod clamp bolt. Or ream out the hole in the tubing and use a ¼ inch nut to hold the tubing to the tripod. Clamp the microphone in the booster clip and tape the mike cord to the tubing.

Some kind of extension clamp is useful for any microphone tests, but essential for nearfield measurements.

TEST PROCEDURES

In each of the following tests, prepare your test leads carefully to avoid adding external resistance to the circuit. Make connections with alligator clips, banana plugs, or spade lugs, as appropriate to the terminal. Avoid twisted wires that can slip loose and remove the load from your audio generator, possibly damaging a sensitive VTVM by voltage surges.

The simple test chassis shown in Fig. 3-1 is recommended for easy hook-ups and quick reference to known resistances. The parts list and layout for this chassis are given in Fig. 3-5. Two comments should be made on the parts list. First, if the power amplifier is used to drive the speaker, increase the 2W resistors to 5W. Secondly, precision resistors are hard to find and very expensive in larger than fractional watt sizes, but you can get a 10-pack of 5%, 10 Ω 2 W resistors for less money. Take the ten resistors to a high school physics laboratory for measurement, or borrow a similar value of precision resistor for comparison. From the ten you should have at least two resistors that measure as close to 10 Ω as a precision 1% resistor.

When completed, the test equipment coupler should be tested for internal resistance.

Test 1: Free Air Resonance

Free air resonance can be tested by too different methods, depending upon the equipment you have on hand. The oscilloscope

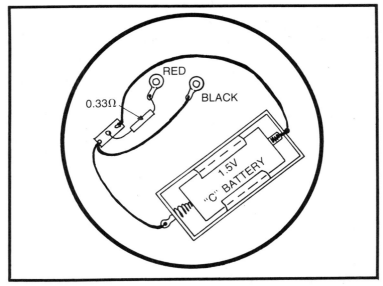

Fig. 3-4. Layout of parts in an empty tuna can for polarity/damper tester shown in photo.

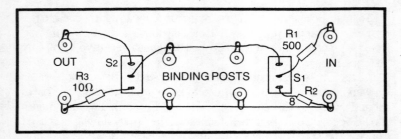

PARTS LIST

NO.	DESCRIPTION	
1	CHASSIS BOX 3" × 6" OR LARGER	
2 PKG.	5-WAY BINDING POSTS	RADIO SHACK 274-661 OR EQ.
1 PKG.	INSULATED ALLIGATOR CLIPS	RADIO SHACK 270-378 OR EQ.
2	SPDT MINI TOGGLE SWITCH	RADIO SHACK 275-326 OR EQ.
1 PKG.	TEST PROBE WIRE (RED)	RADIO SHACK 278-553 OR EQ.
1 PKG.	TEST PROBE WIRE (BLACK)	RADIO SHACK 278-554 OR EQ.
4 PKG.	BANANA PLUGS	RADIO SHACK 274-721 OR EQ.
2	5-10Ω 1% PRECISION RESISTORS	
1	500-1000Ω 2-WATT RESISTOR	
1	8Ω 2-W RESISTOR	

Fig. 3-5. Layout of parts in the test instrument coupler. This is rear view, scale up to fit a 3½" × 6¼" chassis. Coupler in photo was identical to this except for a 3 position switch and extra resistors at S2.

method is especially useful for speakers with a broad impedance peak.

Method I: VTVM. Test Set-Up A (shown in Fig. 3-6). If the speaker diameter is 8 in. or greater, hang it from a cord or strap so that the cone is in a vertical position. Small speakers with lighter cones can be set facing upward on a bench. Keep the space near the cone free from obstructions.

Procedure: Set the VTVM on the 1.5 V scale or a lower one. Turn the switch (S_1) to the speaker pole. Adjust the audio generator output to produce a small deflection on the meter at 200 Hz. Run down the frequency scale until you reach the peak voltage. Locate the exact frequency of the peak and record this as f_s, the free air resonance. If the peak is a broad one, choose the midpoint of your maximum voltage readings. For example, if the peak voltage occurred at 51, 52, and 53 Hz, you would record f_s at 52 Hz. Or if the peak occurred at 38 and 39 Hz, you would record 38.5 Hz.

Repeat the test under different conditions of temperature and humidity if possible. If f_s does change with conditions, make the final test at a similar temperature and humidity to that of the room where the speaker will be used.

Method II: Oscilloscope. Test Set-Up B (shown in Fig. 3-7).
Procedure: Set the audio generator at 200 Hz. The pattern on the scope grid should be an ellipse. Sweep downward in frequency

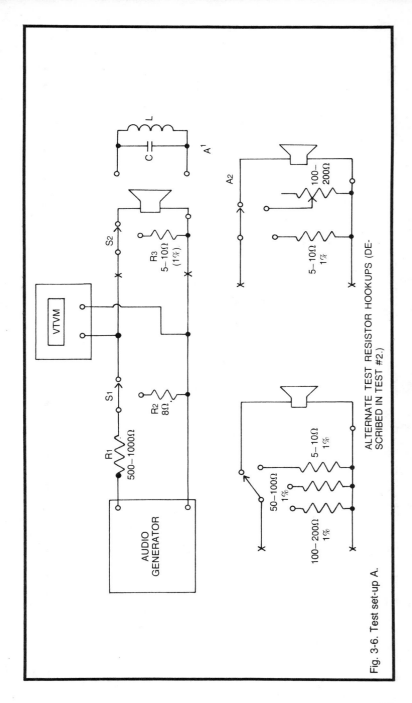

Fig. 3-6. Test set-up A.

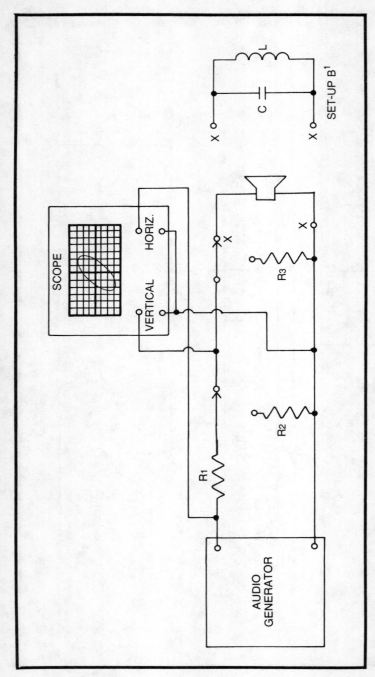

Fig. 3-7. Test set-up B.

until the ellipse rotates to the left and closes into a straight line with maximum vertical deflection. Record frequency on the audio generator as f_s. Figure 3-8 shows the patterns you should see on your oscilloscope at frequencies above and below f_s.

Test 2: Impedance

Both methods for testing impedance require a precision resistor if you plan to measure Q.

Method I: Precision resistor standard. Test Set-Up A.

This test requires a precision resistor or a resistor that has been measured accurately. If you can not find 1% resistors, you can get a 10-pack of 5% resistors and have a high school laboratory measure them on a Wheatstone bridge, or you can borrow a precision resistor for comparison. Out of ten resistors, you should have at least two that have a resistance of 10 Ω within a 1% tolerance. Use one for the coupler and the other as an external test resistor for Test 21. If you cannot get 10 Ω resistors, you can use one of any similar value.

Procedure: Switch the 10 Ω precision resistor into the test circuit. Adjust the generator output to show a "1" near the low end of the VTVM scale. You will assign the value of 10 Ω to this reading. For example, if you have a sensitive VTVM, with a full scale reading of 0.1 V, the audio generator would be set to produce a reading of 0.01 V across the precision resistor.

Switch to the speaker and read the voltage at 10 Hz intervals from the lowest frequency of your audio generator to 100 Hz, at 25 Hz intervals to 200 Hz, at 100 Hz intervals to 1000 Hz, and then at 1000 Hz intervals above 1000 Hz. Record the readings in ohms. In the example above, a reading of 0.02 V would be recorded as 20 Ω impedance; 0.03 V as 30 Ω, and so on. Your readings are in ac voltage, but the 500 Ω series resistor converts the circuit to a constant current and makes the voltage across the speaker proportional to the speaker's impedance, so you can consider the numbers to represent ohms. Do not forget to record the impedance value and frequency of the peak impedance (f_s) in addition to the readings made at regular intervals.

In some tests you may have to use more than one voltage scale, especially if your speaker has a high impedance peak at resonance. Your voltmeter may not give linear readings on adjacent scales, so it is useful to have a switch with at least three positions and an extra precision resistor of 100 to 200Ω (see Set-Up A) to calibrate your higher range readings. The procedure is to make the low impedance readings as described above, but to switch scales and put the higher value precision resistor in the circuit when the voltage rises above

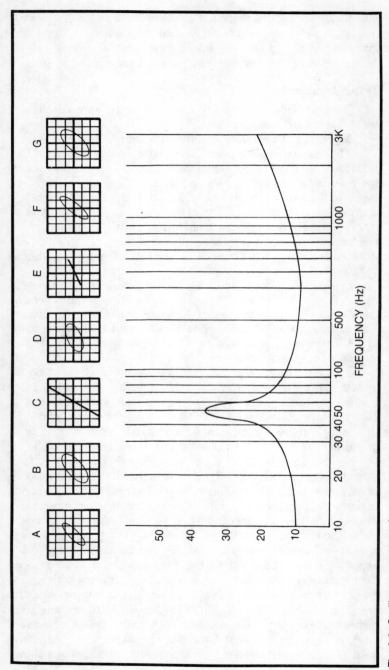

Fig. 3-8. Oscilloscope patterns for various points on an impedance curve.

the scale limit. Adjust the generator to produce an appropriate voltage across the precision resistor on the higher VTVM scale. For example, if you switch from a 0.1 V scale to a 0.3 V scale, and if your precision resistor is 110 Ω, you would set the audio generator output to read 0.11 V. Then make the high scale readings. When you go back to the lower scale, do not forget to recalibrate the generator output to give appropriate direct impedance readings for that scale.

Method II: Variable Resistance. Test Set-Up A_2.

Calibrate a 100 to 200 Ω variable resistor by gluing a long pointer to the control knob and putting a scale on the chassis behind the knob. Use an ohmmeter to measure the resistance at various settings of the control knob, and mark the resistance values in ohms on the scale.

Procedure: Measure the voltage across the speaker's voice coil at each desired frequency and then switch to the potentiometer. Adjust the control knob on the potentiometer to obtain the same voltage across it that you observed when the speaker was in the circuit. Record the value shown on the potentiometer scale. You will also need a precision resistor if you plan to measure Q.

Test 3: Speaker Q

Test Set-Up A.

This measurement is adapted from Dr. Richard Small's method.

If you are using the test instrument coupler, check it for internal resistance according to Test 21 before proceeding.

Procedure: Set VTVM to lowest scale on ohms function. Switch S_1 on coupler to put the 8 Ω resistor in the audio generator circuit, and cut off the audio generator from the rest of the circuit. Turn S_2 to 10Ω precision resistor.

Touch the VTVM leads together and zero the meter pointer. Reconnect to test circuit and adjust the ohms control to read the exact value of your precision resistor. Switch S_2 to speaker. Read the dc resistance of the voice coil and record this as R_e. Switch back to the precision resistor to see if the ohmmeter has drifted. If necessary, readjust the ohms control and make another reading.

Set the audio generator to the frequency of f_s. Read the impedance value and record it as Z_{max}. Calculate the value of r_o from the formula:

$$r_o = \frac{Z_{max}}{R_e}$$

Record the value as r_o.

Find ($\sqrt{r_o}$) and record its value.

Calculate the value of reduced impedance (Z') where:

$$Z' = \sqrt{r_o} \times R_e$$

Run down the frequency scale with the audio generator until you reach the frequency where the speaker's impedance is equal to Z'. Record this frequency as f_1. Then run up the scale from f_s until you find the higher frequency where the impedance is equal to Z'. Record this frequency as f_2. Check the accuracy of your work by this test:

$$f_s = \sqrt{f_1 f_2}$$

The solution of this formula should be accurate within about 1 Hz, or 2%, whichever is greater.

Find the speaker's mechanical Q (Q_{ms}) by:

$$Q_{ms} = \frac{f_s \sqrt{r_o}}{f_1 - f_2}$$

And its electrical Q (Q_{es}) by:

$$Q_{es} = \frac{Q_{ms}}{r_o - 1}$$

The speaker's total Q at f_s is obtained by:

$$Q = \frac{Q_{es} \times Q_{ms}}{Q_{es} + Q_{ms}}$$

Here is an example. Measurements made on an 8 in. woofer show the following values:

$$R_e = 5.8 \ \Omega$$
$$f_s = 54 \text{ Hz}$$
$$Z_{max} = 40 \ \Omega$$

therefore:

$$r_o = \frac{40}{5.8} = 6.9$$

and:

$$\sqrt{r_o} = 2.63$$
$$Z' = 2.63 \times 5.8 = 15.3 \Omega$$

Running up and down the frequency scale, we find f_1 and f_2, which are the frequencies where the impedance is 15.3Ω, to be:

$$f_1 = 40.8$$
$$f_2 = 72.7$$

Note that:

$$\sqrt{40.8 \times 72.7} = 54.5$$

an indication that the frequencies are about right.

$$Q_{ms} = \frac{54 \times 2.63}{32.2} = 4.4$$

And:
$$Q_{es} = \frac{4.4}{5.9}$$
$$= 0.75$$

So the speaker's total Q at f_s is:
$$Q = \frac{0.75 \times 4.4}{0.75 + 4.4}$$
$$= 0.64$$

The Q of a speaker can change with time as the suspension compliance changes, or if the voice coil moves out of proper position. Measured at the time of purchase, one 15 in. woofer had a Q of 0.4. After being stored face down for a few months, the Q measured 0.6. The cone had drifted forward, removing the voice coil from the region of maximum flux density. This contributed to distortion and poor damping. Knowing the proper Q for your speaker can be useful as a check on its health as well as on its enclosure requirements.

Figure 3-9 shows the impedance curves of two 8 in. speakers with test points used for Test 3 marked on the curves.

Test 4: Q Modification

Test Set-Up A (plus small flat baffle and fiberglass bat).

Procedure: Install the speaker on the flat baffle. Measure its Q (Test 3). Cut a 1 in. thick sheet of fiberglass batting large enough to cover the back of the speaker and reach the baffle around the speaker rim. Make a small hole through the fiberglass for the speaker leads and bring the leads out through the hole. Staple the fiberglass to the baffle, covering the back of the speaker. Stretch the fiberglass blanket tightly as you staple it.

Measure the Q again. If Q is lower than desired, try a thinner sheet of fiberglass. If the Q is too high, try a thicker piece.

This test is useful when you want to use a speaker with a Q higher than 0.6 in a ported enclosure. (You can measure Q on a closed box speaker after the speaker is installed.) When you find the right thickness of fiberglass, make a note of it and install the same thickness of fiberglass over the speaker when it is installed in the enclosure.

Example: A 10 in. speaker on a 14 × 14 in. flat baffle had the following Q measurements:
$$Q_{ms} = 3.98$$
$$Q_{es} = 0.55$$
$$Q = 0.48$$

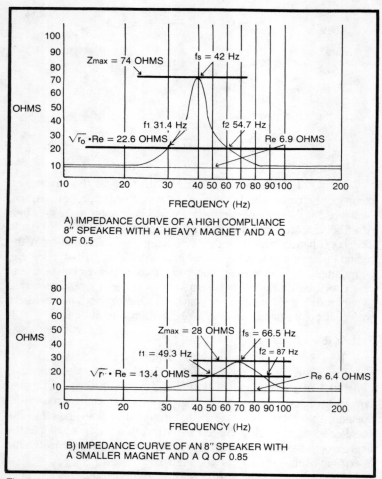

Fig. 3-9. Impedance curves of two speakers showing test data obtained for Test 3.

After a blanket of fiberglass was stretched over the back of the speaker and stapled to the board, the new Q measurements were:

$$Q_{ms} = 1.15$$
$$Q_{es} = 0.53$$
$$Q = 0.36$$

The mechanical damper over the back of the speaker has more effect on the mechanical Q than on the electrical Q, but the electrical behavior of the speaker is affected. For example, the impedance at resonance (Z_{max}) is reduced by the blanket from 31.5 Ω to 12.1 Ω.

The blanket does this by damping the motion of the cone at resonance, which reduces the motional impedance. If Z had been reduced by using a smaller magnet, but having the same motion in the cone, the Q would have been raised instead of lowered. Normally one associates a rise in Q with a lowering of Z_{max} by any cause, but here it is the damped cone motion that changes the Z_{max}, rather than a reduction in magnet strength.

This reduction in Q would permit the speaker to give optimum performance in an enclosure with a cubic volume of less than 1/3 that of the undamped speaker. However, the speaker loses some efficiency when damped by this method.

Test 5: Cone Mass and Compliance

Test Set-Up A or B (plus a known mass of non-drying modeling clay).

Procedure: Find f_s.

Select a lump of modeling clay with an appropriate mass for your speaker as follows:

Speaker Diameter	Mass of Clay
Up to 8 in.	5 g.
8 in.	10 g.
10 in. or larger	20 g.

Have the clay weighed at a school lab or make a balance with a pencil, a ruler, and a nickel as described in the section on "Homemade Test Equipment."

Press the blob of clay onto the speaker cone, preferably near the center. On most speakers, the bumps on the cone where the voice coil leads protrude make a good footing for the clay. As you press down the clay, support the back of the cone with your fingers (Fig. 3-10).

Measure the frequency of resonance again with the mass of clay (M′) on the cone. The new resonance (f_s') will be lower in frequency than f_s because the clay increases the effective mass of the cone. Calculate cone mass by:

$$M = \frac{M'}{(f_s/f_s')^2 - 1} \quad \text{(grams)}$$

Knowing the mass and the free air resonance of the speaker, we can find its compliance by:

$$C_{ms} = \frac{1}{(2\pi f_s)^2 \, M}$$

Fig. 3-10. Support the woofer cone with one hand while you press down a blob of modeling clay with the other.

The 8 in. woofer used for Test 3 had a free air resonance of 54 Hz. When 10 g. of clay are added to the cone, the new resonance is 43.2 Hz. So the cone mass is:

$$M = \frac{10}{(54/43.2)^2 - 1}$$
$$= 17.8 \text{ grams}$$

If we substitute 17.8 for M in the formula for compliance:
$$C_{ms} = 1/[(6.28 \times 54)^2 \times 17.8]$$
$$= 0.48 \times 10^{-6} \text{ cm/dyne}$$

(Comparison is easier if C_{ms} is always stated as a number times 10^{-6} cm/dyne.)

Test 6: V_{as}

If you run the test by both methods and get a minor variation, use the larger value. If there is a large difference, with the calculated V_{as} being considerably greater than that obtained by the box method, check for air leaks in the box, the gasket, or the speaker. Speakers with cloth surrounds can leak through untreated areas on the cloth, so seal these with a thin coat of silicone sealer.

Methods I and II are each subject to estimates, the effective cone area in I, and the volume of the speaker in II. The estimated speaker volumes listed in Fig. 3-2 are higher than the real volumes, but these figures usually give good results. Any leakage can cause the figure to be too low, and so Method III is more reliable.

For a more accurate closed box measurement, you can make the hole in the box just the right size to front mount the speaker from the outside, the way it would be front mounted in its final enclosure. Make a gasket by fitting a strip of plastic tape around the back of the speaker rim where it touches the box hole. Screw the speaker down tightly for the test. Of course you must have the speaker wires installed inside the cabinet by one of the air tight methods described in Chapter 4. Unfortunately, this treatment usually limits the use of the test box to one speaker.

Method I: Mass and Compliance Method. Test Set-Up A or B.

V_{as} is the cubic volume of air that has an acoustical compliance for the speaker that is the equivalent of the cone's mechanical compliance. It is more convenient to use the metric system of measurements for the calculations, but the final volume will be converted to cubic feet.

Procedure: Find the speaker's C_{ms} by Test 5. Measure in centimeters the effective cone diameter by measuring the diameter of the piston area plus the surround on one side only as shown in Fig. 3-11. Calculate effective cone area by:

$$A = \pi r^2.$$

Then: $\qquad V_{as} = C_{ms} \, d \, c^2 \, A^2$

Where:

d = density of air in grams/cm^3 (often represented in physics texts by the Greek letter rho)
c = speed of sound in air (cm/sec)
A = effective piston area of cone in sq. cm.

The density of air varies with humidity and temperature. A typical value, as given by a physics text, would be 0.00129 g./cm^3. The velocity of sound also varies, a typical value being 34400 cm/sec. When used in the formula above, it is convenient to write these values with powers-of-ten notation: 1.29×10^{-3} for d, and $(3.44 \times 10^4)^2$ or 11.8×10^8 for c^2.

For the 8 in. speaker in the previous test, the diameter of the piston area plus the surround on one side was 16.2 cm., so for its radius of 8.1 cm:

$$A = \pi r^2$$
$$= 206 \text{ sq. cm.}$$
$$A^2 = 42436 = 4.24 \times 10^4$$

And substituting the values for A^2 and C_{ms} in the equation for V_{as} in powers of ten notation:

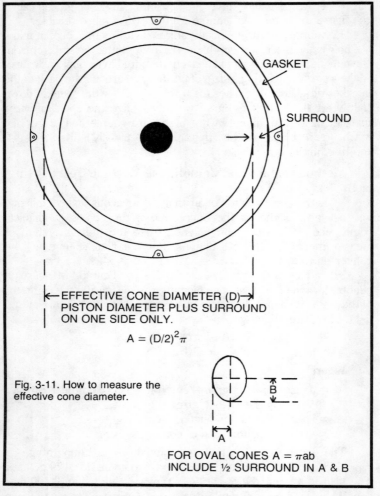

Fig. 3-11. How to measure the effective cone diameter.

$A = (D/2)^2 \pi$

EFFECTIVE CONE DIAMETER (D)
PISTON DIAMETER PLUS SURROUND
ON ONE SIDE ONLY.

FOR OVAL CONES $A = \pi ab$
INCLUDE ½ SURROUND IN A & B

$$V_{as} = (.48 \times 10^{-6})(1.29 \times 10^{-3})(11.8 \times 10^8)(4.24 \times 10^4)$$
$$= 31 \times 10^3 \text{ cm}^3$$

To convert cm³ to in³ use the conversion factor:

$$16.387 \text{ cm}^3 = 1 \text{ in.}^3$$

so:

$$V_{as} \text{ (in}^3\text{)} = 31{,}000 \text{ cm}^3 / 16.387 \text{ cm}^3/\text{in.}^3$$
$$= 1891.7 \text{ in.}^3$$

or:

$$V_{as} \text{ (ft}^3\text{)} = 1892 \text{ in.}^3 / 1728 \text{ in.}^3/\text{ft.}^3$$
$$= 1.1 \text{ ft.}^3$$

This method of finding V_{as} is accurate, except that the rule of thumb for adding half the surround to the cone diameter is simply an estimate of effective cone area. The standard box method automatically takes care of that problem, but requires another kind of estimate.

Method II: Standard Test Box. Test Set-Up A or B, plus test box.

If your speaker has a good front gasket and if the speaker board is smooth, you are ready to test. If not, glue a sheet of smooth rubber, cork, or felt to the top of the test box. After the glue is dry, cut the gasket material out of the speaker opening with a razor blade. Fill any gaps in the speaker's gasket with modeling clay.

Procedure: Center the speaker over the hole in the standard box and apply enough pressure to make a good seal (Fig. 3-12). Measure its frequency of resonance on the test box (f_{ct}). Then calculate V_{as} from:

$$V_{as} = (f_{ct}/f_s)^2 - 1 \times V_B$$

where:

f_{ct} = frequency of resonance on test box
V_B = corrected volume of test box

Since the speaker is outside the box, we must add the speaker's cubic volume to that of the box. Figure 3-2 shows typical values for

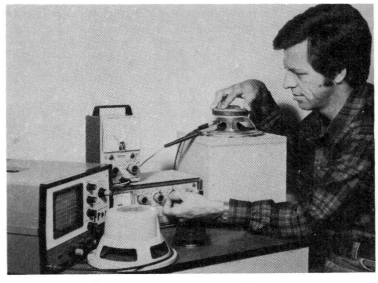

Fig. 3-12. Using a standard text box to measure V_{as}.

the cubic volume of speakers of various sizes and the corrected volume for the test box.

When the previously described 8 in. speaker with a 54 Hz free air resonance was placed on a test box of 0.5 ft.³ internal volume, the new resonance was found to be 88 Hz. When we add the speaker volume (0.15 ft.³ from Fig. 3-2) to the box volume, V_B (corrected) becomes 0.65 ft.³ So:

$$V_{as} = (88/54)^2 - 1 \times 0.65$$
$$= 1.08 \text{ ft.}^3 \text{ or } 1.1 \text{ ft.}^3$$

Method III: Ported Box Method. Test Set-Up A or B (with speaker mounted on or in any ported box and no damping material in the box).

If the port is adjustable, tune the enclosure to approximately f_s, but this step is not mandatory.

Procedure: Set the audio generator to the lowest frequency range. Sweep the range to find the three critical frequencies: that of the lower impedance peak, f_L; that of the upper impedance peak, f_H; and that of the lowest point in the valley between the peaks which is the box tuning frequency, f_B.

Calculate V_{as} from:

$$V_{as} = V_B \times \frac{(f_H + f_B)(f_H - f_B)(f_B + f_L)(f_B - f_L)}{f_H^2 \, f_L^2}$$

A 6 in. × 9 in. speaker was tested in a 1400 in.³ enclosure. Because the speaker was front mounted, it did not occupy interior enclosure space equal to its full volume. If we assume that 10% of the enclosure was occupied by speaker and tuning duct, the net volume would be 1260 in.³ or 0.73 ft.³ The critical frequencies were: f_H, 78.5; f_B, 48.7; f_L, 26.5.

$$V_{as} = 0.73 \times \frac{(88.5 + 48.7)(88.5 - 48.7)(48.7 + 26.5)(48.7 - 26.5)}{(88.5)^2 (26.5)^2}$$

$$= 1.2 \text{ ft.}^3$$

This method has the advantage that minor leaks will not affect the results, but a carefully made standard box (closed) will give just as good results with most speakers.

Test 7: f_B, Tuned Frequency of Ported Box

Which method you use will depend on the equipment you have. If you use both I and II, and the oscilloscope results disagree with the VTVM reading, use the VTVM reading, but try to correct any enclosure or speaker flaws that can produce different values for the two methods.

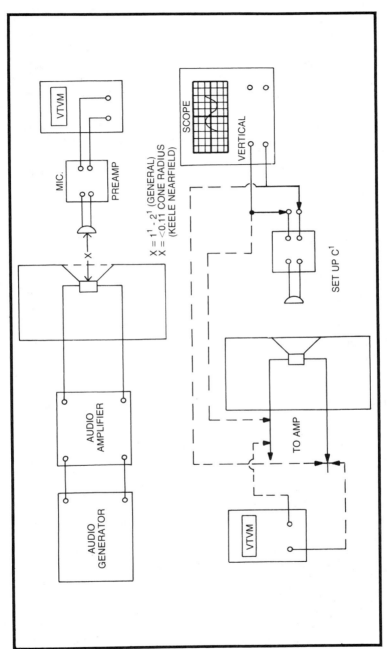

Fig. 3-13. Test set-up C.

Method I: Impedance Method. Test Set-Up A (with speaker in ported box).

Procedure: Set the audio generator on the lowest frequency scale and make an impedance run to locate the three critical frequencies: f_L, the lower frequency peak; f_H, the upper frequency peak; and f_B, the frequency with the lowest value of impedance located between f_L and f_H, which is the tuned frequency of the box. Record the frequency of f_B.

If the valley of the impedance curve is so flat that you cannot identify f_B you can seal the port with the palm of your hand, or screw a board with a gasket over it, and measure the frequency of the system resonance, which will occur at an easily identified single impedance peak. Record this frequency as f_c. Then calculate f_B by:

$$f_B = \sqrt{f_L^2 + f_H^2 - f_c^2}$$

A 6 in. × 9 in. speaker was installed in an 0.8 ft.^{3} box, and the test frequencies were: f_L, 27.5 Hz; f_H, 86 Hz; and, when the port was sealed, f_c, 74 Hz.

$$f_B = \sqrt{(27.5)^2 + (86)^2 - (74)^2} = 51.7 \text{ Hz}$$

Method II: Impedance and Phase Method. Test Set-Up B.

Procedure: Procedure is the same as for Method I except that the three critical frequencies can be identified by both the height of the scope pattern and by the closing of the ellipse to a straight line. This change of pattern makes identification of the box frequency easier if the valley between the peaks in the impedance curve is relatively flat near f_B. If the zero phase condition is obviously different from the maximum and minimum impedance readings, it means that there is a leakage in the system. Check the speaker surround if it is treated cloth.

Method III: Visual Method. Test Set-Up C (without microphone or VTVM).

Procedure: Set audio generator on lowest frequency band. Adjust the output to feed a 5 to 10 V signal to speaker. Adjust a bright light so that you can watch the vibration of the cone while you sweep the frequency range near f_B. At f_B the cone will appear to stop moving.

This is a good test for enclosure sealing as well as for tuning frequency. If there are significant leaks, the cone will continue to vibrate somewhat at all frequencies. Search out and stop all leaks until there is no apparent movement at f_B.

Method IV: Listening Method. Test Set-Up D (shown in Fig. 3-14).

Note that this is a method of finding f_B for a ported enclosure, or for finding the correct duct length to tune a box to a predetermined

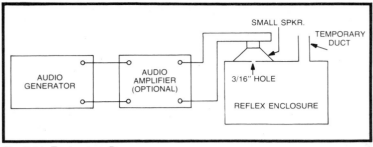

Fig. 3-14. Test set-up D.

f_B without installing the speaker or even without using the speaker that will ultimately be installed in the enclosure. The test speaker can be any small speaker. There must be only two openings in the box, the 3/16 in. hole for the test speaker, and the port. The small hole for the test speaker can be made on the cut-out line for the large speaker hole and be used later for entry by a saber saw; or, if the large speaker hole has already been cut, a board with a 3/16 in. hole can be temporarily screwed to the baffle over the speaker hole. Use a weatherstrip ping gasket to seal the board to the baffle.

Procedure: Set the audio generator on the lowest scale and sweep the frequency range around the expected frequency of f_B. Press down on the small speaker to make firm contact between the speaker gasket and the board. Listen carefully to the sound from the port, but do not block the opening to the port with your head. At f_B you can easily hear a peak in output from the port.

Test 8: Efficiency

Test Set-Up A.

Procedure: Measure f_s, Q_{es}, and V_{as}.

Calculate half-space reference efficiency (usable efficiency when placed against a room wall) by:

$$\text{Eff.} = \frac{2.7 \times 10^{-8} \times f_s^3 \, V_{as}}{Q_{es}}$$

For the first 8 in. speaker used for the test the specification

$$f_s = 54 \text{ Hz}$$
$$Q_{es} = 0.75$$
$$V_{as} = 1.1 \text{ ft.}^3$$

So:

$$\text{Eff.} = \frac{2.7 \times 10^{-8} \times (54)^3 \times 1.1}{0.75}$$
$$= 0.6\%$$

Notice that efficiency is reduced by a heavy cone which requires more energy to drive. A heavy cone will lower f_s. Efficiency is increased by a greater compliance (less mechanical resistance), which raises V_{as}, or by a more powerful magnet which provides more drive and lowers Q_{es}.

Test 9: Frequency Response

If you make only one frequency response test with a microphone, you will undoubtedly know less about your speaker's response that if you use Method III, the listening est.

Method I: Microphone Tests. Test Set-Up C.

Mount the microphone on a stand or clamp. To minimize room effects, place the microphone about a foot or so from the speaker, and in line with the mid-range and tweeter if you are testing a multi-speaker system.

If you live in a quiet neighborhood, you can avoid room effects by making outdoor measurements. Choose a day with no wind and face the speaker into a large open space with no reflecting surface directly behind it. Set the speaker on a stand, 1 ft. to 2 ft. above the ground.

Procedure: Set your audio generator at 1000 Hz and adjust the output level to produce a useful deflection in the VTVM pointer, such as 0 dB, or a midpoint on a voltage scale. Run up and down the frequency scale and record the frequency and dB, or the voltage of each peak and valley. If you find that 1000 Hz represented a peak or valley, choose a frequency with a more typical reading as a reference.

Move the speaker and the mike to another location in the room, but maintain exactly the same distance between them for a second run. If your VTVM is not calibrated in dB's, use the table in the appendix to convert to dB's the ratio of each voltage reading to the reading at 1000 Hz. Plot the results on log graph paper. In plotting curves you can smooth out the minor wiggles. Any difference between the two curves will show changing room effects, not speaker variation.

To test the speaker's dispersion, rotate the enclosure so that the speaker will fire at a 30 degree angle to the line from the microphone to the speaker. Check the frequency response at 1000 Hz intervals and compare it to the results obtained with the microphone directly in front of the speaker.

After making two or three readings at one mike distance, make a run with a slightly different distance and record the results. Compare the results with your earlier readings. A microphone can help you choose a grille cloth. Make readings with the grille cloth between mike and speaker; then remove the grille cloth and repeat.

Fig. 3-15. Measuring the frequency response of a woofer by Keele's nearfield microphone technique, Test 9, Method II.

Do not be discouraged if you find sharp peaks or valleys in the response curves, especially below 500 Hz, and do not expect much information from the curves except as a rough guide for matching drivers, or to investigate the effects of any enclosure or wiring changes. If you find a bad peak, you can use the method described in Chapter 14 to remove it.

Method II: Nearfield Microphone Technique. Test Set-Up C (with microphone extension).

The Keele nearfield pressure method of measuring low frequency response can be used in any acoustical environment. Nearfield measurements make use of the high pressure near the bass driver before it is reduced by low room pressure. This method works about as well as a dead room for frequencies below that where the radius of the cone is less than a quarter wavelength of sound.

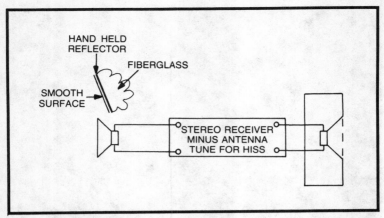

Fig. 3-16. Test set-up E.

The microphone should have a flat response down to the lowest frequency to be measured. Some inexpensive dynamic microphones work well down to 50 Hz or even lower; others fall off rapidly below 100 Hz. A fall-off in your tests could be your speaker—or your mike.

Procedure: Place the microphone as close as possible to the center of the cone, within 0.11 of the radious. For example, for an 8 in. speaker with an effective cone diameter of 6.5 in. and a radious of 3.25 in., the distance should be less than ⅜ in. from the mike to the cone.

Nearfield pressure is directly proportional to farfield pressure, so you can use the nearfield method to compare bass response of speakers and enclosures.

To measure port output of reflex systems, place the microphone at the center of the vent, flush with its front edge. Your meter or scope should show the response of the vent rising to a peak at the tuned frequency of the box (f_B).

Method III: Listening Tests. Test Set-Up D (minus microphone and VTVM shown in Fig. 3-13).

Procedure: Set your audio generator at 1000 Hz. Reduce the generator output until you can barely hear the tone. Sweep through the frequency range, noting points where the output peaks or drops out. (Variatons in loudness levels are easier to detect at low volume.)

Discount any drop-off below about 200 Hz; this will occur in most acoustic situations, and your ear is less sensitive here.

Test 10: Distortion and Power Handling

Both Methods I & II require an oscilloscope.
Method I: Oscilloscope with Microphone. Test Set-Up C'.

Procedure: Set audio generator at frequencies to be tested: chiefly 100 Hz and below, where distortion is greatest.

Connect the scope vertical input leads to amplifier output, and check waveform for distortion. This hook-up should show a pure sine wave.

Switch vertical leads to the output of the microphone preamplifier. Any change in waveform indicates distortion, either in the speaker or the microphone. By testing several speakers, you may be able to verify whether the distortion is in the microphone or the speaker. Extra ripples show harmonics of the fundamental tone (Fig. 3-17A).

If the sine wave is pure, increase the amplifier output until distortion appears. Connect a VTVM across the amplifier output and measure voltage. Calculate the undistorted power output of speaker by:

$$P = \frac{E^2 \text{ (volts)}}{Z \text{ of speaker}}$$

For example, an 8 in. speaker shows distortion at 60 Hz when the drive to the speaker equals 8.4 V. The speaker's impedance at 60 Hz is 10 Ω. So its power handling ability at 60 Hz is:

$$P = \frac{(8.4)^2}{10} = 7 \text{ W}$$

This test should be used with care to avoid damaging speakers or amplifiers.

Sweep the audio generator upward in frequency, and look for sub-harmonics. A sub-harmonic is evident as an undulating envelope of a high frequency wave train (Fig. 3-17B). This kind of distortion may be found in cheap speakers with light, thin cones.

If you do not have an oscilloscope, you can learn to identify distortion by ear. Listen to a low frequency tone at low volume, then increase the drive until you can hear a second, higher pitched tone.

Method II: Oscilloscope without Microphone. Test Set-Up B (with an amplifier driving speaker through the series resistor).

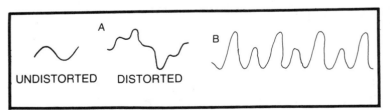

Fig. 3-17. Test 10 waveforms.

Procedure: Set audio generator at the lowest frequency band.

Explore the low frequency band and notice if the scope pattern is a clean pattern. Most speakers will show a pure pattern: a straight line, a circle, or an ellipse at low volume. At some point, however, the ellipse will flatten or loop, or the straight line will bend. These effects show distortion.

Apply the power test as in Method I.

NOTE: Make sure the amplifier you are using is free from distortion at the output level needed by monitoring the signal from the amplifier as in Set-Up C'.

Test 11: Peak Displacement Volume of Cone, V_D

Test Set-Up:

Audio generator and amplifier driving unbaffled speaker.

Procedure: Hold speaker to your ear and listen to the nearfield distortion while sweeping the 25 to 60 Hz range with a constant drive voltage to the speaker. Increase drive voltage until there is a sudden, noticeable increase in distortion which you will hear as a change in tone. Estimate the linear displacement of the cone at this point. You can do this by the phenomenon known as persistence of vision. When you have established the level at which you hear distortion at a certain frequency, move the speaker into your field of vision and focus your eyes on some part of the cone, or on the flexible leads to the voice coil where they attach to the cone. As the cone vibrates, you will see a ghost image of its peak displacement. Make an estimate of this distance and record it as the peak linear displacement of cone, X_{max}. Calculate the peak displacement volume by:

$$V_D = A \times X_{max}.$$

Where:

A is the effective cone area.

For example, an 8 in. speaker with an effective cone area of 30 in.^{2} appears to have a peak displacement of 1/16 in. Its peak displacement volume is:

$$V_D = 30 \text{ in.}^2 \times 1/16 \text{ in.} = 1.875 \text{ or } 1.9 \text{ in.}^3$$

The quantity V_D is useful in calculating minimum port area in a reflex enclosure.

Test 12: White Noise Test

Test Set-Up E

Prepare a small (approximately 1 ft.^{2}) reflector with a smooth hard surface on one side. Glue a thick (2 in. to 4 in.) pad of fiberglass to the other side.

Procedure: Set the FM receiver on monaural. Connect the speakers that you want to compare to each channel and turn the balance control all the way left or right. Disconnect the antenna and tune for no signal, just intercarrier hiss.

Hold the unbaffled speaker near your ear. Hold the reflector behind the speaker with its smooth side toward the rear of the cone. Move the reflector in and out of the speaker's rear field as you listen to any change in sound. The reflector will probably produce a nasal quality by making the frequency response more peaky in the mid-range, unless the speaker is so peaky that you cannot hear a difference.

Reverse the reflector so that the fiberglass pad faces the rear of the speaker. Repeat the procedure above. Then reverse the reflector, listening for any change that occurs as you turn from the fiberglass side to the smooth side (Fig. 3-18). As you learn to recognize reflective coloration, you will appreciate the need for damping material behind all enclosed cones.

Install one speaker, or set of speakers, in its enclosure and compare the sound from the enclosed speaker to that of the unbaffled speaker. If you hear a change in the mid or high frequency color, you will know it is caused by the box. Use this test to find the best amount and placement of damping material in the enclosure to reduce mid-range peaks.

Fig. 3-18. To interpret the White Noise Test 12, first practice with a smooth reflector and a damping pad behind a bare speaker.

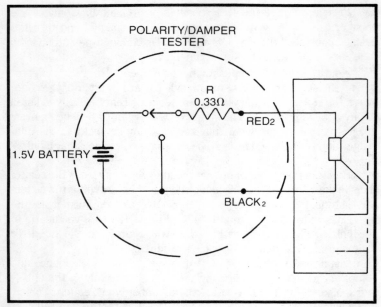

Fig. 3-19. Test set-up F.

Test 13: Polarity

You can test polarity with an oscilloscope, a home-rigged set-up, or your ears.

Method I: Homemade Tester for Speaker Polarity. Test Set-Up F (shown in Fig. 3-19).

Procedure: Connect the leads from the tester to the speaker (Fig. 3-20). Throw the switch to connect the battery to the speaker, and observe the cone movement. Reverse the leads if necessary to make the speaker cone move outward at the circuit "make" and inward on the "break." Then mark the speaker terminal to which the positive (red) lead is connected with a red dot or a plus sign.

Method II: Newspaper Test for Speakers behind Grille Coth. Test Set-Up F.

Procedure: Follow the procedure for Method I, but hold a single sheet of newspaper close to the grille cloth so that it covers the front of the speaker. Have someone operate the battery tester while you watch the movement of the newspaper. Reverse the leads of the tester and compare. Mark the terminal as above.

Method III: Oscilloscope Test. Test Set-Up G (shown in Fig. 3-21).

Procedure: Place the two microphones (cheap crystal lapel mikes work well for this) side by side at equal distance from a single cone speaker (Fig. 3-22). Feed a low frequency signal to the speaker while you adjust the scope controls, until you get a line that is inclined by 45 degrees to the right or left of the vertical axis of the screen. With identical mikes, the line should be inclined to the right, indicating in-phase signals, but even identical mikes may be wired out of phase. It is not important which way the mikes are wired, but the direction of the line will indicate an in-phase speaker signal because you are testing a single speaker that is obviously in phase with itself. Make a note of the slope direction.

Next, place the microphones before any pair of speakers you want to test. Make sure that both mikes are located at exactly the same distance from the speaker each one is monitoring (Figs. 3-23 and 24). If the speakers are in phase, the line on the screen will be inclined the same direction as that for the single cone speaker. If they are out of phase, it will be rotated 90 degrees to the left or right.

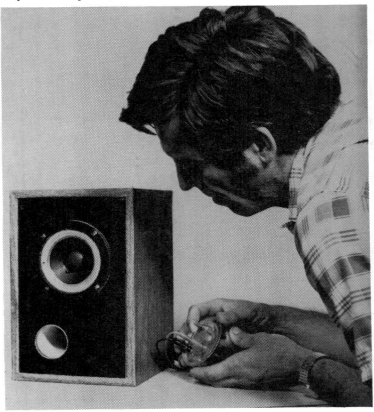

Fig. 3-20. Tests 13 and 19 can be made with a simple polarity/damper tester diagrammed in Test Set-up F. (Project 15 being tested.)

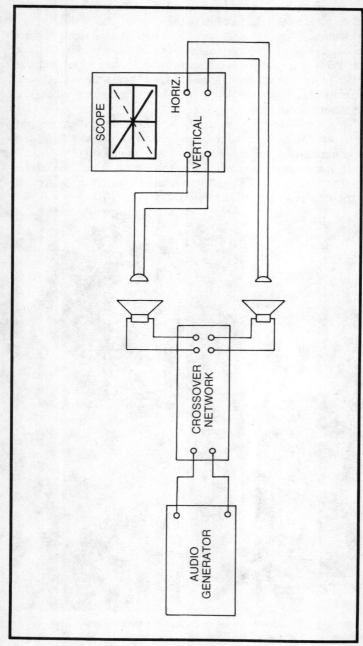

Fig. 3-21. Test set-up G.

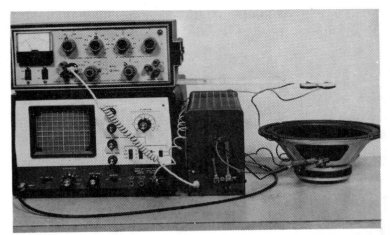

Fig. 3-22. First step in using microphones to phase speakers is to test the phasing of the microphones.

Method IV: Listening Tests for Stereo Speaker Systems. (Test set-up: speakers connected to receiver or amplifier.)

Procedure: Place the two speakers close together, face to face. Feed a 100 Hz tone to the amplifier from a test record, audio generator, or from music with heavy bass tones. Reverse the leads to one speaker. The correct hook-up will have more bass response.

Test 14: Speaker Phasing in Crossover Network

Method I is not practical for high frequencies because the wavelength of the sound is too short; even a minor error in distance from microphone to speaker will alter the phase.

Method I: Oscilloscope Test. Test Set-Up G (Use for woofer/mid-range or for woofer/tweeter phasing).

Fig. 3-23. Microphones must be equidistant from woofer and tweeter. Tweeter should be elevated so voice is at level of woofer's voice coil.

91

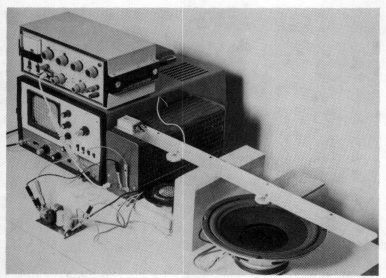

Fig. 3-24. Phasing test of crossover network by microphones and oscilloscope, Test 13, Method III. (This is another view of Fig. 3-23.)

Procedure: Set the audio generator at the crossover frequency of the two drivers. Proceed as in Method III of Test 13.

A circle instead of a line on the scope indicates that the speakers are out-of-phase by 90 or 270 degrees. This occurs with any simple 6 dB/octave network. In this case, wire the speakers with similar terminals (plus or minus) connected to the same side of the line in the crossover network. For 12 dB/octave 3-way networks (2 network components per driver), wire the tweeter with the same polarity as the woofer, and the mid-range driver with reversed polarity.

Method II: Microphone Test. Test Set-Up C.

Procedure: Set the audio generator at one octave below the crossover frequency (f/2) and run up the scale to an octave above the crossover point (2f). Record the voltages at 100 Hz intervals. Note values of dips.

Move the microphone to various distances from the speakers and run the test again, recording voltages.

Reverse the leads to one driver and repeat the above tests. Choose the hook-up that produces the highest average voltages.

Test 15: Voice Coil Inductance

Voice coil inductance is not the same at all frequencies because it depends partly on the acoustic properties of the speaker, as well as on the number of turns of wire, the diameter of the coil, and the iron around it. Method I gives an approximation of voice coil inductance, which is useful in crossover network design. The advantage to Method II is that you can investigate the inductive reactance at the crossover frequency. Note that you can also use Test 16 to measure voice coil inductance.

Method I: VTVM Test. Test Set-Up A.

Procedure: Set the audio generator at f_s. Sweep upward in frequency until you reach the frequency with the minimum impedance. Record the impedance value. Then sweep upward again until the impedance is twice the lowest value. Record the frequency. If you assume that the increase in impedance is produced by voice coil inductance, then at the frequency where $Z = 2 Z$ minimum, then $X_L = Z$ minimum and $L = X_L/2\pi f$.

For example, an 8 in. woofer has a low impedance of 5 Ω at 200 Hz. Its impedance is twice that (10 Ω) at 730 Hz, so X_L at 730 = 5 Ω.

And:

$$L = \frac{5}{2\pi 730} = .001 \text{ H or } 1 \text{ mH}.$$

Method II: Oscilloscope Test. Test Set-Up B.

Procedure: Set the audio generator at f_s, then sweep upward until the ellipse on the scope closes into a straight line. This is the minimum impedance level, where the speaker's capacitive reactance equals its inductive reactance. Adjust the scope controls for equal vertical and horizontal deflection, so that the line is inclined exactly at 45 degrees and centered on the CRT.

Switch the precision resistor into the circuit. Adjust the output of the audio generator until the vertical deflection is 1 cm. for a 10 Ω resistor.

Switch back to the speaker. Set the audio generator at the desired crossover frequency. Measure the impedance value from the vertical deflection on the grid (Fig. 3-25). Record this as Z. Measure the sine of the phase angle by:

Sine $\theta = A/B$

The reactive component of the impedance will be:

$X_L =$ Sine θZ

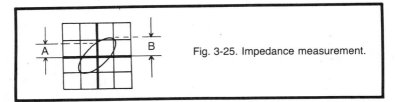

Fig. 3-25. Impedance measurement.

For example, the same 8 in. woofer used in the example for Method I was tested by the oscilloscope method. This speaker was to be used with a crossover frequency of 2000 Hz. Tests showed:

$$Z \text{ at } 2 \text{ kHz} = 18 \ \Omega$$
$$\text{Sine } \theta = 13/18$$

Because we calibrated the oscilloscope with the 10Ω resistor, there is no need here to do any arithmetic with the sine function. The reactive component of the impedance is 13 Ω.

So:

$$L = \frac{X_L}{2\pi f}$$
$$= \frac{13}{2\pi 2000}$$
$$= 0.001 \text{H or } 1 \text{ mH}.$$

In this case the value was the same for both methods, probably by chance.

Test 16 Inductance of a Choke Coil

Methods I and II can be used to measure a speaker's voice coil inductance, but unless you have a large-value paper, oil filled, or mylar capacitor on hand, the frequency will be far above the useful crossover frequency. A speaker does not have constant voice coil inductance at all frequencies; otherwise, this method is a way to measure inductance accurately. The oscilloscope in Method II will show resonance much more clearly than the VTVM in Method I.

Method I: VTVM Test. Test Set Up A' (Substitute a coil and parallel capacitor for the speaker in set-up A).

Use a high quality paper or mylar capacitor, if possible. Electrolytics may need a dc voltage to perform to their full capacitance; they can be slightly inductive, and they are more variable in quality.

Procedure: Try the various frequency bands on your audio generator until the voltage across the coil and capacitor increases sharply to a maximum. This maximum voltage at resonance can be as much as 100 times the voltage of $f_{/10}$. Label the frequency of resonance (f) and calculate the coil's inductance by:

$$L = \frac{1}{(2\pi)^2 \ f^2 \ C}$$

Note: You can substitute 40 for $(2\pi)^2$.

A 1μf capacitor was connected across an unknown coil and the frequency of resonance was 3450 Hz.

$$L = \frac{1}{(40)(3450)^2(0.000001)}$$

$$= 0.0021 \text{ H or } 2.1 \text{ mH}$$

Method II: Oscilloscope Test. Test Set-Up B' (Substitute coil and parallel capacitor for speaker in set up B).

Use a high quality paper or mylar capacitor.

Procedure: Try the various frequency bands on your audio generator until the ellipse on the scope becomes a straight line, indicating resonance. Calculate the inductance from:

$$L = \frac{1}{4\pi^2 f^2 C}$$

For example, a 2.1 µf capacitor was connected across an unknown coil and the frequency of resonance was 5400 Hz.

$$L = \frac{1}{(40)(5400)^2(0.0000021)}$$
$$= 0.4 \text{ mH}.$$

Method III: VTVM, (No parallel capacitor.) Test Set-Up A (Substitute coil for speaker).

Procedure: Switch VTVM to "ohms" function and measure the dc resistance of the coil. Switch the precision resistor into the circuit and calibrate the audio generator output level so that the VTVM shows a reading equal to the value of the resistor.

Switch the coil into the test circuit again. Sweep the audio generator up the frequency scale until the impedance of the coil is great enough to make an accurate reading. Record the value of the impedance (Z) and the frequency (f). Calculate the inductive reactance of the coil at that frequency by:

$$X_L = \sqrt{Z^2 - R^2}$$

NOTE: If R is insignificant, say 1 Ω or less, ignore it and assume that $X_L = Z$.

Calculate inductance from:

$$L = \frac{X_L}{2\pi f}$$

For example, a large air-core coil had an impedance of 10 Ω at 323 Hz. So:

$$L = \frac{10}{(6.28)(323)}$$

$$= 0.0049 \text{ H or } 4.9 \text{ mH}.$$

Test 17: Matching Inductance to Capacitance

Test Set-Up H (shown in Fig. 3-26).

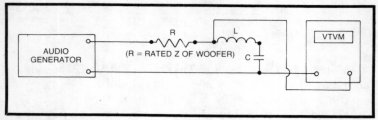

Fig. 3-26. Test set-up H.

Procedure: Select R equal to Z of speaker. Set the audio generator at a frequency well below the desired crossover frequency and sweep upward in frequency until the voltage across the capacitor and inductor falls to a minimum. This is the crossover frequency for any speaker that has an impedance in ohms equal to the value of R.

If the null frequency is too low, remove some turns from the coil and recheck until resonance occurs at the desired frequency. If it is too high, solder a piece of magnet wire to the end of the coil wire and add some turns.

This test is useful for crossover networks only if the woofer and the tweeter have approximately the same impedance at crossover, and if they are equal in output over the near frequency band.

This test can be used for any filter for which it is desirable to make the inductive reactance equal the capacitive reactance of a predetermined value at a certain frequency.

Test 18: Crossover Network Response

Test Set-Up I (shown in Fig. 3-27).

Procedure: Connect a VTVM or a VOM across a low frequency terminating resistor. A VOM is supposed to work better here because it is isolated from the power line, but a VTVM works satisfactorily.

Run a frequency response test for the low frequency channel.

Connect the VTVM or VOM across the high frequency terminating resistor.

Run a frequency response test for the high frequency channel.

Plot the frequency response of the two channels on a sheet of log graph paper. If the curves for the legs of the crossover leave a large gap near the crossover frequency, replace the terminating resistors with others having double the resistance of the originals. Repeat the test. If the curves from the first test show overlapping at the crossover frequency, replace the resistors with others having half the original value and repeat the test.

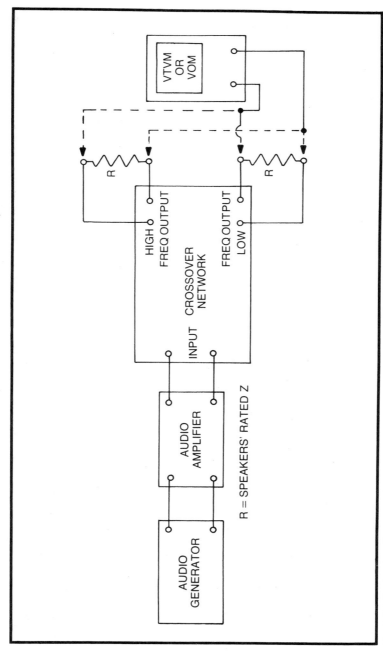

Fig. 3-27. Test set-up I.

This test is useful for identifying the crossover frequency and for proper load matching for an unlabeled crossover network.

Test 19: Speaker Damping

Test Set-Up F.

NOTE: The 0.33 Ω resistor shown in the homemade tester circuit is about right for a speaker that will be used with an amplifier that has a damping factor of 25 at the speaker's resonance. Some amplifiers have a rated damping factor much higher than that, but advertised damping factors are based on tests at 1000 Hz. It is usually lower at low frequencies. If you know your amplifier's damping factor, you can calculate the optimum value for R by:

$$R = \frac{\text{rated Z of speaker}}{\text{amplifier damping factor}}$$

Procedure: Install the speaker in its enclosure. Connect the leads from the tester. Flip the switch on and off. If the damping is inadequate, you will hear a thud or bong instead of a click. When you add damping material, the sound at the circuit "make" will clear up before the sound at the "break." If putting more damping material in the box doesn't do an adequate job, staple a fiberglass blanket over the back of the speaker. A thickness of about 1 in. is usually enough, depending on the density of the fiberglass.

This test is particularly useful for speakers in ported boxes where you can't easily measure Q of the finished product, but it can be used for any kind of system if you don't have the equipment for measuring Q.

Make the final adjustment of damping material by a listening test.

Test 20: Critical Listening

Test Set-Up:

Speakers connected to amplifier or receiver.

Choose program material that you know, if possible. Voice is good for judging naturalness. An outdoor test on a quiet day is especially valuable for comparing two speakers, because indoors even a slight difference in position can produce a totally different room effect. One exception is the omni-directional speaker, which sounds very different in open air because there are no reflections.

Procedure: Adjust the volume control for low sound level. If the speaker is peaky, you may hear only the peaks, making it sound thin or limited in range. This can be a rigorous test.

Don't be too dejected on the basis of a single program source. Commercial recordings and radio station transmissions vary greatly in audio quality.

In some ways, this is the most important test of all. If the other tests say "go," but your ears say "no," change it. Speaker design should be based on technical data, but there is still an art to getting the best sound out of a speaker system.

Test 21: Coupler and Lead Resistance

Test Set-Up A (with extra precision resistor in place of speaker).

Procedure: Switch VTVM to the same low voltage ac scale used for testing speakers.

Switch S_2 to a 10 Ω precision resistor.

Set audio generator at 100 Hz. Adjust the output to produce a significant deflection on the VTVM: 0.1 V on the 0.3 V scale, or 1.0 V on the 1.5 V scale, or any other value that can represent a 10 Ω load directly as 10.

Switch S_2 to the external precision resistor. Note any difference in the reading on the VTVM. Make a correction in your readings to account for any added resistance in the coupler or test leads.

For example, if the reading with the internal 10 Ω resistor is 0.10 V and, with the external 10 Ω precision resistor 0.105 V, you should subtract a ½ Ω correction to all impedance readings to compensate for the resistance in the coupler and test leads.

To verify the resistance, switch S_1 to an 8 Ω resistor and set the VTVM on "ohms" function, lowest scale. Switch S_2 to internal resistor.

Touch the leads of the ohmmeter together to zero meter, then reconnect. Adjust the ohms control to read exactly 10 Ω. Switch S_2 to the speaker circuit with an external 10 Ω precision resistor connected to the leads. Any difference in the two resistance readings will show the added resistance of the test coupler and leads.

A Box for Your Speaker

Anyone who has ever listened to an unbaffled speaker knows how necessary an enclosure is for high fidelity sound. Few listeners can judge a speaker unless it is installed in a suitable baffle, or box. I once watched some people who were comparing two unbaffled speakers, a 6 in. by 9 in. high compliance model with a heavy magnet, and a cheap 5 in. by 7 in. stiff-coned speaker with a thimble-sized piece of iron. Their verdict was unanimously for the cheap speaker. Its high Q and middle bass resonance gave the impression of its having some low frequency response, while the lack of a baffle killed the low frequency response of the better speaker.

To see why a bare speaker sounds bad, consider Fig. 4-1. The plus signs represent an increase in pressure as the cone moves against the air; the minus signs, a decrease (Fig. 4-1A). When some of the air from the convex side of the cone meets the air from the concave side, it reduces the pressure difference and cancels the sound to the degree that the mixture has occured. At high frequencies, the sound is increasingly directional, so there is a little cancellation, but low frequencies are omni-directional. At low frequencies, where the wavelength is long in relation to the diameter of the cone, the waves can curve back around the cone so that the out-of-phase waves mix (Fig. 4-1B). So it is that in free air, a speaker "sees" an acoustical short circuit below a certain critical frequency that rolls off its bass response, making it sound thin.

You can lower this critical frequency by installing the speaker on a flat board, or baffle. The new critical frequency will be reached when the path from the rear of the cone to the front of the cone equals one-half the wavelength of the sound. If you want to use a flat baffle, there is no point in lowering the critical frequency below the driver's frequency of resonance; responce falls off rapidly below that point anyway. Even so, for most woofers, a suitable flat baffle would be large. For a speaker with a 40 Hz resonance, the baffle should be 14 ft. in diameter; that figure should indicate why flat baffles have never been very popular.

At one time many high fidelity experts considered the "infinite baffle" to be the most perfect way to install a speaker. A real infinite baffle would be so large that the out-of-phase back wave would never reach the front of the cone. Such a baffle would load the cone equally on both sides and produce no air volume resonances. A room wall can act as an infinite baffle down to very low frequencies, often to 20 Hz or below, but there are practical problems. You must have a house (or a tolerant landlord), you must avoid a cavity at the front of the speaker, and there is always danger to the speaker cone if someone shuts a door in either of two rooms.

The description of what happens to an unbaffled versus a baffled speaker may seem to indicate that sound cancellation is total below a certain frequency; what happens is acutally more complex than that. At any frequency where the phase difference is equal to the original difference (180 degrees) *and* where the intensity of the two sounds are equal, complete cancellation occurs. The phase difference may be other than 180 degrees; at some frequencies, the reflected sound from one side of the cone may be in phase with the sound from the other side of the cone, and there is reinforcement instead of cancella-

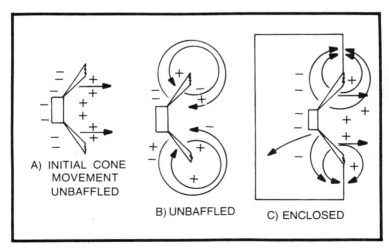

Fig. 4-1. Why an unbaffled cone is inefficient at low frequencies.

tion. The net result of all this in- and out-of-phase mixing is an uneven response, as well as a dropping low frequency performance.

A baffle or enclosure does more for a speaker than merely preventing cancellation. Any device that receives or transmits power must be properly matched to its source, or to the medium in which it operates. A speaker which receives power from the amplifier and delivers power to the room air should be correctly matched to the electrical output impedance of the amplifier, and to the acoustical impedance of the room. Its electrical impedance varies with the frequency of the signal, but we can try to avoid gross mismatching. At the acoustical end of the speaker's role, the problem is more complicated.

When an unbaffled speaker tries to pump air around, it meets little resistance from the air, so, like a car engine piston out of its cylinder, it can do little work. A large cone meets more resistance than a small cone, so it can push more air. Any speaker, large or small, can work better if it is installed in a baffle or box so that the air cannot get away so easily when the cone moves against it. Various kinds of enclosures serve this function to various degrees, depending upon their design. A horn, which flares from a small cross-sectional area near the driver to a large mouth, does the best job of coupling the cone to the air. Its gradual transition from the increased pressure at the cone to the lower pressure in the room raises the radiation resistance of the speaker, so the speaker can convert electrical energy into sound with high efficiency. Unfortunately, low frequency horns must be extremely large.

Damping the speaker's resonance is yet another function that different kinds of enclosures fulfill to different degrees. When a speaker cone vibrates out of control, it has "hangover." A speaker with hangover reproduces music about as well as a muddy rear view mirror on your car reflects the world around you. Hangover kills definition, or the ability of the listener to identify individual tones and instruments. The speaker continues to vibrate after the signal from the amplifier ends, mixing tones no longer present in the signal with new tones. If you have a piano handy, you can hear the sound of hangover by pressing the right pedal while you run your finger up and down the keys. Although the right pedal is called the damper pedal, it really removes the damping pads from the piano strings when it is depressed. Without the pads against them, the strings vibrate long after the hammers strike them.

Many speakers have a transient response that is somewhere between that of a damped and a undamped piano. Although these speakers do not completely muddle the sound, they do not respond accurately to pulses; they vibrate just long enough after the signal ends to blur sharp transients. Such a speaker's resonance is likely to be excited by other bass tones, causing one-note bass of the kind produced by old-time juke boxes. Casual listeners sometimes like the effect until they become accustomed to true bass that moves up and down the scale.

When the speaker is installed in a box, the air pressure against the cone adds an acoustical resistive load to it. Like the damping pads in the piano, the air pressure damps excessive cone movement. In the complete system, two kinds of damping affect the cone: electrical damping and acoustical damping. A large magnet and a modern amplifier with low output resistance will provide the necessary electrical damping, but acoustical damping helps reduce distortion. Speaker distortion is most troublesome at low frequencies, so any speaker enclosure that provides good low frequency damping offers a useful feature.

An unbaffled speaker has almost no mechanical damping, except for whatever damping may be applied by the suspension system. This lack of damping is apparent from the speaker's impedance curve, which reflects the degree of cone movement at various frequencies. To highlight this, when we measure impedance, we remove some of the electrical damping by putting a resistor in series with the speaker.

Putting the speaker in a closed box changes the resonance in two ways: it moves the resonance up in frequency, but down in amplitude. The pressure behind the speaker offers some control on the cone that would have moved more freely in open air.

Labyrinths, reflex boxes, and horns provide more damping effect at low frequencies. The damping effect of the quarter wave labyrinth usually occurs at the most restricted range of frequencies. The reflex is similar in action, but horns damp the cone over a wider range. Impedance curves of speakers in each kind of box show the damping effect. A reflex box produces the familiar double humped curve, but a horn produces a smooth, almost constant, impedance line.

We will later examine in detail how these different enclosures vary in design and performance. The important points to observe now are that all of them serve the same kind of functions to varying degree, and that all of them are subject to the same general rules of construction.

DESIRABLE ENCLOSURE CHARACTERISTICS

The word *enclosure* imples some kind of box, although engineers may call it a *baffle*. The engineers called the early closed box speaker enclosures *infinite baffles* because these boxes completely isolated the backwave from the front. Both terms have largely passed out of use, and today we hear *enclosure, cabinet*, or just *box*.

In addition to serving the acoustical functions mentioned earlier, an enclosure must meet at least one more requirement if it is to be used in a high fidelity system: it should not add its own voice to the sound. This rule is often overlooked by advertising claims for low-fi and even for some high fidelity equipment. One current speaker line uses a special lumber core for its baffle board which, the company's advertising claims, gives the speakers a "rich tonal quality." Another ad boasts that a certain table model radio has a wood cabinet that gives it the "same mellow resonance that grandma's big console used to have." Besides suggesting that resonance is a good thing, this ad promotes a fallacy that goes back to the beginning of the era of reproduced sound. It is rare in contemporary advertising only because modern low-fi sets have little or no wood in them.

The idea that wood made the best speaker boxes probably did reach its peak acceptance in the days of grandma's big console; but it started much earlier, with the art of violin making in the seventeenth century. In the piano or violin factory what kind of wood used made a big difference—a well recognized fact. But a speaker system is not a musical instrument. An instrument is designed to *produce* a sound of a specific quality by adding certain overtones (actually harmonic distortion) to the fundamental tone. A musical instrument that did not do this would produce an uninteresting sine wave, like an audio generator. The speaker, however, must be able to *reproduce* a sine wave or a complex wave on command; if it adds anything to the amplifier's signal, it is producing distortion. The enclosure walls must not vibrate. Such vibration adds coloration to the sound and absorbs energy, weakening the speaker's bass response. Even if the panels produce some low frequency response, it will be at various phase angles to that of the cone's response, and cause interference.

Enclosures do produce useful sound, but only by air volume or air column resonance. A properly designed reflex or labyrinth enclosure augments the bass response of the woofer by this kind of resonance. Note this difference: an air resonance can be controlled, either by volume or dimensions, but panel vibration is more unpredictable.

ENCLOSURE MATERIALS

Wood is one of many materials that can be used for speaker enclosures. It is convenient to use and attractive when finished. However, any material that is opaque to sound, rigid, and reasonably well damped is satisfactory. Industrial grade chipboard, of the quality better grade cabinet makers use, is an excellent choice, having greater density than wood and being well damped. As a rule of thumb, you can rate the quality of materials by their density: the higher the density, the better the material. Concrete and brick have been used for large, high quality enclosures, but they require a good foundation under the speaker system and so are not a practical choice for most of us. In the case of steel, which is poorly damped, density is not a good guide. Ceramic tile, either drain tile or flue tile, provides a ready-made enclosure of four walls that is satisfactorily rigid. Tile has the disadvantage of weight and a limited choice of shapes.

An old rule of thumb suggests that all speaker enclosures should have walls made from ¾ in. material. This is the minimum thickness for floor models, but the thickness of the walls should be appropiate to the size of the box. Sub-miniature boxes for 5 in. speakers can be made with walls of ½ in. plywood and be as rigid as larger enclosures with 1 in. walls. Here is a rough guide for minimum wall thickness: for speakers up to 6 in. in rated diameter, ½ in. material; for 8 to 12 in. speakers, ¾ in. material; and for 15 in. or larger speakers, 1 in. material. Remember that the size of the box has more to do with the necessary thickness than the size of the speaker, but the two usually go together.

Many authorities tell you to brace the walls of your speaker cabinet, but few tell you exactly what happens when you do. A sub-heading in an electronics magazine advises the reader to brace the walls to "lower cabinet resonance." Braces do lower the amplitude of resonance; but, unlike the common view, they also raise the frequency of resonance. Mechanical wall resonance is produced by a mass vibrating against a compliance, as in a speaker. A brace adds stiffness and reduces compliance, so it increases the frequency of resonance. That's good. The higher the resonance, the more easily it can be damped by fiberglass or by other similar material in the cabinet.

What happens to the speaker board does indeed lower its frequency of resonance: when you cut a hole for the speaker, you increase its compliance; when you bolt on the speaker, you increase its mass. The speaker frame will increase the stiffness again, but how much it will do so depends on the rigidity of the frame. The only redeeming virtue of all this is that the speaker's board resonance becomes different from that of the back, and staggered resonances are usually more acceptable than closely grouped resonances.

Most recommendations on bracing a cabinet suggest putting braces around the mid-section of the box, across the short dimension of each panel. Mathematical analyses have sometimes been used to support such bracing. What these studies prove is that a short brace is more rigid than a long brace; but, since the object here is to make the *panels* more rigid, the brace should go in the direction

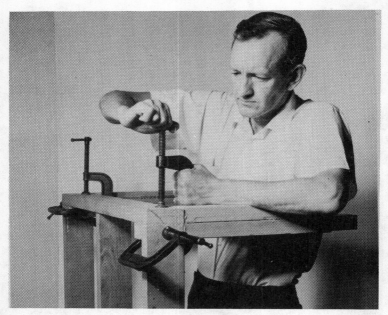

Fig. 4-2. One way to build a large enclosure: Make a frame like a house frame with 2 × 4 framing lumber, then glue, screw, or nail sheets of chipboard or plywood to the frame.

that divides the panel into the narrowest subpanels. The best choice is usually a lengthwise brace for each panel. Experiments by Peter Tappan in the early 1960's showed the effectiveness of this kind of bracing, but it is rarely used.

One way to get non-resonant panels for speaker enclosures is to laminate several materials with different characteristics. You can bond asphalted roofing felt to plywood or chipboard to improve damping, or make a sandwich of three materials, such as a handboard/celotex/plywood sandwich. G.A. Briggs once suggested a hardboard/styrofoam/plywood laminate for a lightweight resonance-free material. Styrofoam has a high ratio of stiffness to mass, but it cannot be used alone because it is almost transparent to low frequencies. Different materials transmit different frequency bands of sound, so a laminate is more likely to be acoustically opaque at all frequencies, in addition to having a broader resonance.

One of the easiest ways to build up an enclosure with laminated panels is to make a frame, like a small house frame, and apply the inside layer to the frame (Fig. 4-2). Then glue and nail the other materials to the box, using the most attractive material on the outside.

TOOLS AND CONSTRUCTION TECHNIQUES

You can build good speaker enclosures with a few hand tools if you have the time and patience to do careful work. Paul Klipsch fitted the parts of his first Klipschorns with just a hand saw and a wood rasp, and the X-3 model Klipschorn that he built that way in 1942 is still working. That model Klipschorn is virtually the same speaker that is produced today with 59 specially shaped pieces of wood. If Klipsch could build one of his complicated horns with hand tools, there is no reason why anyone can't make a simple box without power equipment.

You can even build a box without a saw if your lumber dealer will cut the pieces to a specified size for you, though there may be a charge for that service. Some building material dealers rent special purpose tools.

Here is a list of tools, all desirable, some necessary: carpenter's hammer, screwdriver, tape measure or folding rule, large carpenter's square or try square, hand drill, keyhole saw for cabinet making, a soldering iron or gun, and pliers (diagonal and long-nosed) for wiring the speakers.

A few power tools will save time and effort: a table saw for making dadoed or beveled joints, a saber saw for speaker holes, an electric drill for making pilot holes for screws, and a sander for finishing work.

If you don't have a complete line of tools, you can use ring-shank or screw-thread nails and glue, especially for small boxes. Just make sure the joints fit closely, and your cabinet will have the necessary strength. Screws are better than nails to hold parts under pressure while the glue sets. The most useful size for ¾ in. material is the 1¼ in. length. This length of screw can be used to fasten together two ¾ in. thick parts without penetrating the cabinet exterior. A 1¼ in. wood screw will put a ½ in. of threads in the second layer of wood, enough to hold 35 lbs. in soft wood, and up to about 50 lbs. in hardwood. To use #8 screws, you should drill an 11/64 in. shank hole through the first member, and a smaller pilot hole in the second member. For hard wood, make the pilot hole 3/32 in.; for soft wood and chipboard, 5/64 in.

When you start putting parts together, you may find that it is difficult to get the pilot holes to match the locations of the shank holes. One easy way to locate glue blocks, cleats, or panels for drilling is to drill the shank holes first, then nail the parts together, but do not drive the nails all the way down. Drill the pilot holes in the lower member. You can then pull out the nails far enough to break down the temporary assembly. Leave the nails in the top piece so that they protrude far enough below it to find the previously made

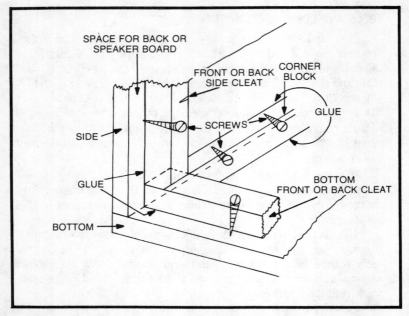

Fig. 4-3. Good corner construction for large enclosures.

holes in the second, then coat the matching surfaces with glue and nail the parts together. Now install the screws; you won't have to fight the pieces to keep them from slipping around as the screws go in. Choose nails that will not penetrate the second piece; "sheet rock nails" are just the right length (1¼ in.) for ¾ in. material.

When screws are used to hold parts together for gluing, they should be spaced at 4 in. to 6 in. intervals. For more pressure, use clamps. Or, if you have no clamps, use ropes or straps to hold the panels together while the glue sets. This works particularly well for small enclosures with 45 degree bevel joints.

The best way to find the exact point where matching parts fit together is to use the parts themselves to measure their dimensions, rather than to mark every cut with a tape measure. The cleats are fitted around the interior of four sides to receive screws from a front or back panel (Fig. 4-3). It is easier to mount the cleats on the panels before putting the panels together (Fig. 4-4). To place the cleats correctly, first install the end cleats, then set the parts together in a temporary box while you mark the outline of the end cleats on the side panels. These marks show the proper length and position of the side cleats so you can avoid large gaps. Fill small gaps with silicone rubber sealant, or latex caulking compound.

Many speaker enclosures are made with a removable back for speaker installation. You can build a better, more rigid enclosure by planning to install the speakers from outside the box. With front mounting, you can glue every joint and follow up with caulking compound. This kind of construction virtually eliminates the chance of air leaks. Want to tune a reflex? Do it from outside the box. Get some heavy cardboard tubes such as rug makers use for shipping rolls of carpeting. Make a hole in the speaker board for the tube. After tuning the box, cut the tube to the right length and glue it into the board so it's flush at the front of the board.

The kind of joints you make at the corners of your cabinet will depend on what tools you have available. With hand tools you are limited, but simple butt joints can be very good if they are close fitted, and if you run glue blocks along each boundary (Fig. 4-5). For plywood, a butt joint has the disadvantage that the end grain shows for the full thickness of the top panel. If you have access to a power

Fig. 4-4. Good corner construction with screws and glue blocks.

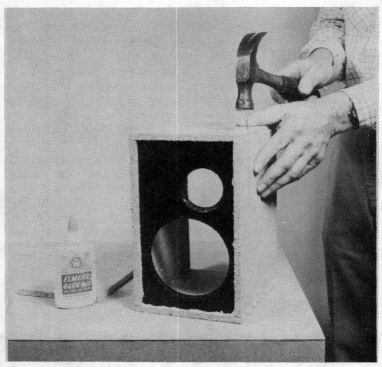

Fig. 4-5. Miniature boxes with butt joints can be put together with glue and nails (Project 17).

saw, you can use the dado blades to make a rabbet joint. This kind of joint shows less end grain, but is usually weaker than the plain butted joint. A 45 degree beveled joint will show no end grain (Fig. 4-6). For the greatest strength, a 45 degree joint is splined, or cut to fit in what is called a lock-mitre joint, but these adaptations of the beveled joint require milling machines.

To maintain an airtight box, the speaker cable from the amplifier can be fed into the cabinet either by tight-fitting bolts, or through a hole filled with silicone rubber caulking compound. If the cable is used to carry the signal through the cabinet wall, it should be knotted on the inside of the box to protect it. When bolts are used, solder the wires to solder lugs on each side of the panel, under the heads and nuts of the bolts. The most convenient connection is a terminal pad with screw terminals. For this, drill ¼ in. holes to permit the wires to go through and to make a space for the screws when they are tightened on the speaker cable spade lugs. Then prepare a piece of lamp cord long enough to be attached to the

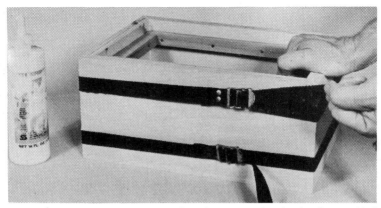

Fig. 4-6. Mitered corner joints on small enclosures can be glued and assembled with straps or ropes to hold them under pressure until the glue sets (Project 5).

terminal on the back or bottom of the box, and yet permit speaker removal without damaging the cord. Bring the lamp cord ends from the inside of the box out through the holes, and solder them to the lugs on the terminal pad. Install the terminal pad with glue from outside the box. Fill the holes on the inside with caulking compound, as shown in Fig. 4-7. This makes a neat installation and helps prevent any shorting of speaker leads.

HOW ENCLOSURE SHAPE AFFECTS SOUND

Most speaker enclosures are box shaped, having six sides with the opposite sides parallel. Such cabinets can be handsome, but this

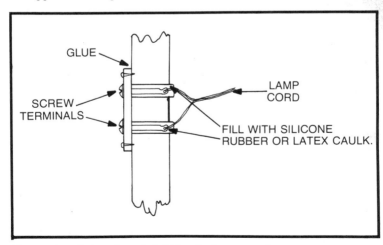

Fig. 4-7. Installation of an air tight terminal on an enclosure.

111

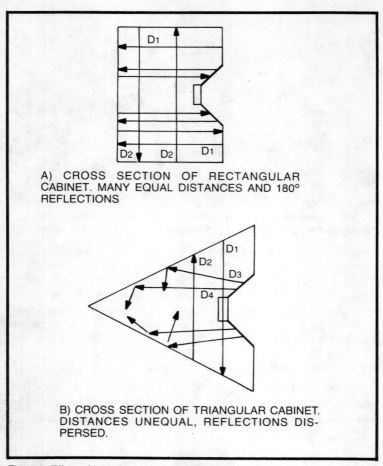

Fig. 4-8. Effect of box shape on internal reflections.

beauty may only be skin deep: parallel walls bounce the sound between them like an echo chamber. Damping material in the box can absorb much of these reflections, but prevention is still worth more than a cure. A box cross section where the sides fit at angles other than 90 degrees is better for preventing internal reflections, as shown in Fig. 4-8.

The external shape of a simple box does about as much damage to a sound wave as do the internal reflections. In one way it is worse, because the problem can not be cured by adding damping material. Sharp edges and projections at the outside corners cause diffraction as the waves reach those corners, and the diffracted waves overlap with destructive interference, particularly in the mid-range. The ideal external shape would be a box that curved away from the speaker, at least in the horizontal plane. A 3-sided front is a good compromise.

In some ways the shape that favors good bass reproduction does not favor treble dispersion. Experiments by the BBC have shown that the narrower the speaker enclosure, the wider the dispersion, especially for mid-range sound. However, a narrow enclosure lets the low frequencies curl back around the box, reducing radiation resistance. A wide base helps to form the wave and to force it forward into a more restricted angle. Another way to limit the angle of radiation is to place the speaker either at the boundary of two plane surfaces, on the floor near a wall, or at the boundary of three plane surfaces, in a corner. In fact, the location of a speaker system in a room usually has much more effect on its bass response than does the shape of the box.

THE FIBONACCI SERIES AND ENCLOSURE DESIGN

Regardless of theory, most speaker enclosures look like simple boxes and will probably continue to be simple boxes for two reasons: ease of construction and appearance. To make the most of these parallel-walled demons, you should avoid extreme shapes, particularly those having any one dimension more than three times any other, and those having a height that is greater than twice the width. Thiele suggests that speaker enclosure dimensions should be made similar to preferred room dimensions for a spread of standing wave modes. He mentions dimension ratios of 0.6 : 1 : 1.6 and 0.8 : 1 : 1.25. The second ratio produces a deep, squat box that is hard to fit into the usual available space.

Thiele's first ratio closely fits what artists sometimes call the "golden mean" or "golden rectangle," which is often attributed to a fifteenth century mathematician, Lucas Pacioli. Actually, the ratio is much older than that. Leonardo Fibonacci, an earlier mathematician, brought it back from the Middle East in the twelfth century. Going back much further in history, the ratio of 1:6 or, more accurately, 1:618, is found in the construction of the Great Pyramid in Egypt. Rounded off to two places, the complete "golden" ratio for a three dimensional object is 0.62 : 1.00 : 1.62. If you look back to Thiele's preferred speaker enclosure ratios, you will come to an astounding conclusion: that a ratio that has been recognized for thousands of years as a preferred set of proportions almost exactly matches that recommended in the late twentieth century to give the best acoustical performance for high fidelity stereo speakers.

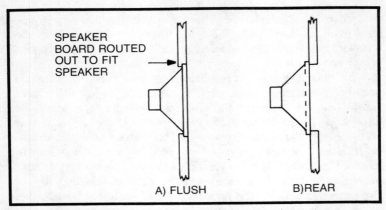

Fig. 4-9. Flush speaker mounting (a) and rear speaker mounting (b).

HOW TO USE THE "GOLDEN RATIO"

First, find the required cubic volume of your enclosure using Chapter 6 for a closed box design, or Chapter 7 for a ported box. If you find the right volume in cubic feet, convert the figure to cubic inches. Then, using a pocket calculator or log tables, find the cube root of the volume. This value should be the internal width of the box. Multiply it by 0.62 to find the internal depth, and by 1.62 to find the internal height.

For example, let's say the required volume is 2600 in.^{3} The cube root of 2600 is 13.75, so the internal width should be 13 3/4 in. The internal height should be 1.62 × 13.75 or 22¼ in.; the depth 0.62 × 13.75 or 8½ in.

If you want to make a cabinet in which the external appearance matches the golden mean, find the right cubic volume and convert it to inches. Take the cube root of this figure, but let it represent the *external* width. Multiply it by 1.62 to get the external height. Now subtract the total wall thickness from each dimension, which would be 1.5 in. for ¾ in. material, to obtain the internal width and height. Multiply these figures, then divide the cubic volume by the product of the internal width and height to get the internal depth.

For the same 2600 in.^{3} enclosure with a "golden ratio" front, we would make the external width 13¾ in. and the external height 22¼ in. If the box material is ¾ in. thick, this would make the internal dimensions 12¼ in. × 20¾ in. The volume 2600 in.^{3} divided by 254 yields a required depth of about 10¼ in.

HOW TO INSTALL SPEAKERS

The ideal position for a speaker cone in relation to the baffle is a flush mounting (Fig. 4-9A). The traditional positioning of the cone behind the speaker board (Fig. 4-9B) can produce cavity resonance as well as reflection and diffraction from the sharp edge of the circular opening in the board. All of these effects produce dips and peaks in the frequency response curve, usually in th mid or high frequency range where the ear is most sensitive to them. Front mounting (Fig. 10A) eliminates the cavity at the front of the cone, but leaves the speaker frame and its gasket projecting from the board. To countersink the speaker frame into the board, you would need a router. A substitute method of flush-mounting can be used by laminating the speaker board and cutting the speaker hole on the outside board large enough to accept the entire speaker frame inside it, as shown in Fig. 4-10B.

Few speakers have the proper gaskets for front mounting. You can coat the speaker board with silicone rubber compound to make an airtight gasket, but removal will be more difficult. One or two layers of black plastic tape on the back of the speaker frame will work fairly well, or you can use thin weatherstripping material.

One of the best ways to fasten the speaker to its board is to use T-nuts (Fig. 4-11). Drill holes through the board to match the location of each speaker frame hole. For T-nut installation, this hole should be one drill size larger than the bolt size. Drive the T-nuts into the holes on the opposite side of the board from the speaker. Use lockwashers under bolts with slotted heads (machine screws) so you can tighten them with a screwdriver. The sharp projections on the T-nuts will hold them while you tighten the bolts from the speaker side of the board (Fig. 4-12). Most hardware stores stock T-nuts in 3/16 in. and ¼ in. sizes.

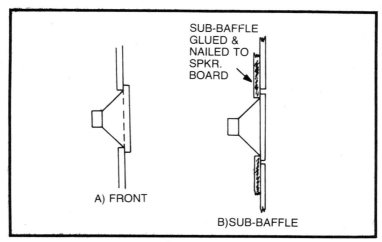

Fig. 4-10. Front speaker mounting A and sub-baffle speaker mounting B.

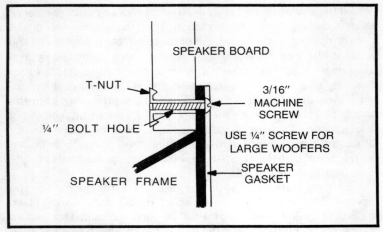

Fig. 4-11. Use of T-nuts.

You can mount small speakers with panhead sheet metal screws. Drill 5/64 in. guide holes in the baffle for #8 screws. Regardless of which method you use to mount your speakers, don't forget to respect the law of the screwdriver: "The chances of putting the screwdriver blade through the cone varies directly with the price of the speaker and inversely with that of the screwdriver." It sometimes helps you to control the screwdriver, especially one with a worn blade, if you square off its blade with a file, but do this *away* from the speaker, and wipe off all the filings. When you are installing a front-mounted speaker, you can place a small board over it for protection while you drive in screws.

SPEAKER PLACEMENT ON BAFFLE

The main rule on driver placement is to put the higher frequency speakers above the lower frequency speakers. In most locations, this will put the tweeters nearer to ear level, where the highs can be heard and will sound more natural. It also puts the woofer closer to the floor for bass reinforcement, although this is not always necessary.

Speakers that cover adjacent frequency bands should be located in a vertical or nearly vertical line with little horizontal displacement. If they were put side by side, there would be a difference in path length from the two speakers to the ear of the listener, and at frequencies where the speakers' output overlaps, this difference would produce phase distortion.

If possible, speakers should be located off the center line of the enclosure. With compact enclosures, the woofer must usually be on the center line for lack of space, but the mid-range and tweeters can be installed off center to make the horizontal distance to each side of the enclosure different. Since this mounting will usually bend the area of coverage slightly toward one side of the enclosure, the second speaker of a stereo pair should be a mirror image of the first. An easy way to lay out the second speaker board is to cut out the holes in the first board, then lay it face down on the second and mark the speaker openings. When the speakers are placed in the room, the left channel cabinet should have the tweeter nearer the right edge (as seen by the listener) and the right channel cabinet should have its tweeter nearer the left. This arrangement will give better listening area stereo coverage, unless the speakers are located very close to each other. In that case, try reversing them to find the best stereo effect.

Fig. 4-12. Driving T-nuts into back of speaker board to receive 3/16 in. machine screws (Project 10).

Fig. 4-13. Where speakers are to be front mounted, make a cardboard pattern that fits the speaker. This is especially useful for oval speakers, as shown (Project 10).

DAMPING MATERIAL

One audio fan once gave a colorful explanation of why damping material is necessary: he said that all the sound from the rear of the speaker cone is distorted and must be absorbed. That is not true, of course, but in most speaker enclosures, the backwave does cause considerable trouble. When the backwave is reflected to the speaker cone, it can cause great variations from an otherwise flat frequency response. The worst effects occur in the mid-range, where the speaker will sound excessively bright. Some listeners like this brightness, considering it to have the elusive quality of "presence;" but after a long listening session, the reaction usually changes to "hard to take." There is an old wives' tale that putting too much damping material in a box kills the speaker's high frequency response. If this is true, it's better dead. However, the tale is really impossible. Almost all speaker systems have tweeters with sealed backs. Reducing mid-range peaks makes the sound smoother, but many people interpret the reduction in peakiness to be a high cut.

Too much damping material can affect response in another way: it can lower the Q of the speaker at resonance, reducing bass response. Cheap speakers with a high Q can benefit from more damping material than good ones; compact boxes can benefit more than large boxes (Fig. 4-14). One effect of loosely filling a box with stuffing is that the fibers absorb heat from the sound energy in the box and give it up again, changing the air action from adiabatic (constant heat) to isothermal (constant temperature) (Fig. 4-15A). The result? A reduction in the speed of sound, and shorter wavelengths that make the box "act" bigger than it is. You can easily add 20% cubic volume this way; theoretically you can add up to 40%.

Don't put damping material in or near the port of a reflex unless you have a special reason to do so (Fig. 4-15B). A reflex port can act with full efficiency and apply its beneficial damping action on the driver only if the port is clear and smooth. Reflex boxes are often larger than closed boxes and need damping material only on the walls near the speaker. The most effective location for the damping material is over the back of the speaker so that the sound will be forced to travel through the blanket of material twice to get back to the cone. A sheet of damping material stretched over the frame can lower the speaker's Q (Fig. 4-15C), but if that is undesirable, the material can be hung like a curtain behind the speaker.

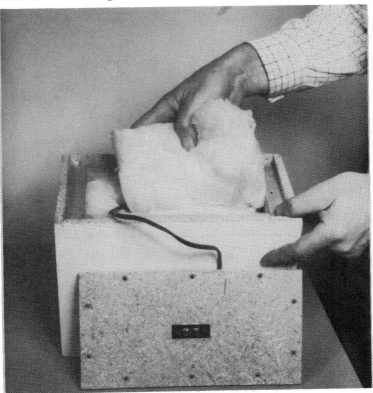

Fig. 4-14. Most miniature closed boxes should be filled with damping material. Note terminal strip on back (Project 17).

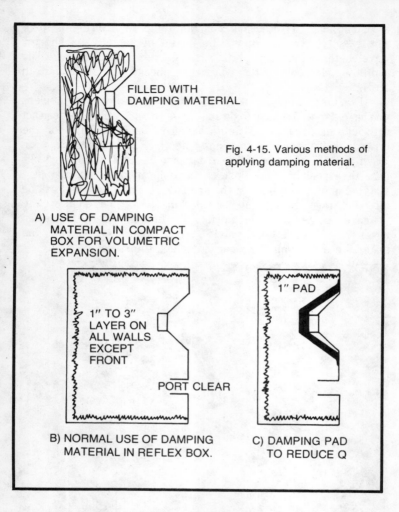

Fig. 4-15. Various methods of applying damping material.

Most speaker manufacturers use fiberglass because its characteristics are well known. Fiberglass is available from hi-fi dealers or electronics stores or, if you need a lot of it, you will find large rolls of house insulation grade fiberglass at building supply stores. The house insulation batting should be pulled away from its paper backing, and it is a good idea to cover this kind of fiberglass with cheesecloth to keep fibers out of the woofer's voice coil.

Felt is more effective at low frequencies. There is a special grade for speaker enclosures, but you can use felt rug padding. Long-fiber wool is excellent, but hard to find. Some transmission line builders claim that dacron batting, which is available from most department stores, is second to long-fiber wool. Other materials have been used from time to time, but most of them have serious disadvantages. The easiest materials to find and use are felt, dacron, and fiberglass.

GRILLE CLOTH

The ideal grille cloth would never sag or stretch, and would be visually opaque but acoustically transparent. Like the perfect loudspeaker, the perfect grille cloth has not been developed. Since manufacturers of grille cloth do not give out frequency response tests, the best way to choose a grille cloth is to hold it up to the light and try to look through it. The most transparent cloth will usually be the best choice for acoustic performance. Good grille cloth is often expensive, so for a low cost system you may have to compromise and use cheap decorator burlap, particularly if you are building an enclosure with a wraparound grille. Grille cloth can damp the port of a reflex enclosure, reducing speaker damping.

FINISHING YOUR CABINET

This may seem to be the final step in building a speaker system, but it shouldn't be. Do your finishing work before installing speakers,

Fig. 4-16. Sculptured foam grille, available from Radio Shack, gives professional appearance to homemade speaker system (Project 4).

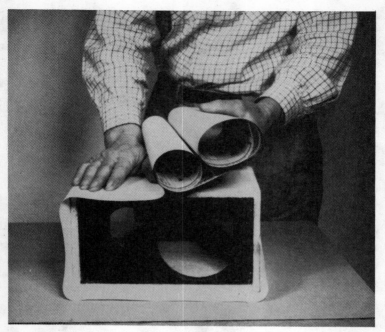

Fig. 4-17. The easiest way to finish chipboard is to use self-adhesive plastic veneer (Project 17).

and keep the speakers away from metal filings or dust. A friend of mine once built a cabinet for a new speaker and, after hearing the speaker, decided that its cabinet should have a furniture quality finish. He stained and lacquered the wood, rubbing the lacquer down with fine steel wool between each coat. Finally, he moved the high gloss cabinet back up to his living room, hooked up the speaker to his hi-fi system, and sat back to listen. Instead of the rich sound he had heard earlier, his new speaker sounded terrible. Inspection showed that when he had rubbed down the finish with steel wool, the speaker's magnet had drawn the airborne iron fibers through the grille and into the gap between the voice coil and the pole piece. End of speaker.

In putting a finish on your speaker enclosures, you can be creative and come up with a unique effect, or you can stick to the conventional plywood and varnish. If you have an imaginative turn of mind, you can use almost any kind of material on the cabinet exterior: paint, fancy wallpaper, cork, ceramic tile, or you-name-it. The easiest way to cover a box is to buy a roll of self adhesive plastic veneer in any pattern that appeals to you (Fig. 4-17). For a more durable finish, get a countertop material like formica.

Some builders have used chipboard with a clear varnish and made no apologies; but, for most people, the prestige exterior is still hardwood plywood, stained and varnished or lacquered. If that is what you want to do, go to a paint store and choose a *single* brand of stain, sealer, and final varnish or lacquer. Get all the printed advice the company puts out on how to apply its products, and follow directions. In this way you will have selected compatible materials.

Instead of going through the steps on a conventional job, here are two unusual methods that give a "close to wood" finish that some people find more natural than a coat of varnish.

First, there is the oiled finish as suggested by JBL. Sand the cabinet with 6/0 sandpaper until it is smooth. Clean it carefully with a soft rag to remove all the sand grit and dust. Then coat it with a mixture of 3 parts boiled linseed oil and 1 part pure gum turpentine, applying this finish with a rag soaked in the mixture. After a half hour, wipe the surface with a clean dry cloth. Sand lightly with 360 grit wet or dry sandpaper. Repeat this process until 3 coats of oil finish have been applied. When the last coat has penetrated, rub it down with soft dry rags. You should renew the finish once or twice a year for the first couple of years. If you do it right, the cabinet wood will grow richer with each coat.

Now for the easiest and cheapest finish of all—shoe polish. By mixing different brands and colors, you can mimic many traditional wood finishes, but do your experimenting on a piece of wood similar to the wood in your cabinet, rather than on the cabinet itself.

Here is a good mixture for changing birch plywood to a medium tone walnut. Stain the wood with a fruitwood stain, using a soft rag to make the stain go on evenly. Let the stain dry overnight. Follow this with a brown shoe polish for a reddish walnut, or midnight brown for the traditional walnut. With a dry rag, wipe off the excess polish while it is fresh. The best job I have ever seen was done with a fruitwood stain, Kiwi "midnight brown" shoe polish, and a Cavalier "otter" boot polish. I don't know if all those materials are available now, but you can probably find similar ones. Finish up with a neutral shoe polish. A big advantage of this method is that you can correct for mistakes without damaging the wood: if you get too much of one color, you can tone it down with a rag dipped in paint thinner. And, if you scratch your cabinet, you know where to find the right scratch filler. Shoe polish is the foolproof furniture finish.

HOW TO IMPROVE A CHEAP COMMERCIAL ENCLOSURE

No one should buy a cheap speaker with the intention of improving it unless he has no tools to make his own, or can get it at a giveaway price. If you have one, here's how to improve it.

Begin by looking at the back of the cabinet. If you can see the speaker, it has to be a cheap one. Replace it. Remove it from the cabinet and take it to an electronics store to match frame size and bolt holes with a high compliance model of the same size.

Check the cabinet construction. Most cheap boxes do not have adequate wall thickness; some have their speaker boards stapled here and there to the sides, or to short blocks. You can glue in cleats for the speaker board and braces for the walls. Add glue blocks to every boundary by coating them with glue and nailing them in place. Put in enough damping material to suppress the internal reflections. You can sometimes make a significant improvement in a cheap speaker just by doing the detail work the factory left out, but don't expect a miracle.

You can judge the construction quality of any enclosure quickly with the knuckle test. Rap each panel sharply. If it sounds hollow or drummy, it needs more bracing. It should sound dead, and the higher the pitch for a certain size panel, the better. Rap near a corner to see how the two sides brace each other. Then check the middle of those same sides. The speaker board and back will usually have the poorest response to this test unless they are extremely well braced.

5 Types of Enclosures

There are four types of speaker enclosures in general use: closed boxes, ported boxes, labyrinths and their variations, and horns. A fifth type, the flat baffle, is found occasionally.

CLOSED BOX ENCLOSURES

The builders of early infinite baffle enclosures had to make them extremely large to maintain good bass response, the reason being that the high fidelity speakers then available were 12 in. or 15 in. models with relatively stiff cone suspensions. The closed box put the stiffness of the air in the box in series with the cone stiffness, raising the resonant frequency of the speaker. Because the free air resonant frequencies were relatively high, often from 50 to 75 Hz, the speaker/box combination produced a frequency response curve that began to roll off well above the bottom of the musical instrument range. If the box were to be large enough to avoid raising the speaker's resonance too much, it had to be big and have a volume of at least 8 to 10 cubic feet.

Then, in the 1950's, loudspeaker engineers achieved a breakthrough in lowering the space requirements for closed box speakers. The people who made the first "acoustic suspension" speaker made a woofer with such a high compliance (low stiffness) that its resonant frequency fell below the reproduced audio band. The woofer had such a compliant suspension that it could easily be driven into distortion, or even damaged, in a large box. However, in a small box, the low compliance (high stiffness) of the air in the box provided the

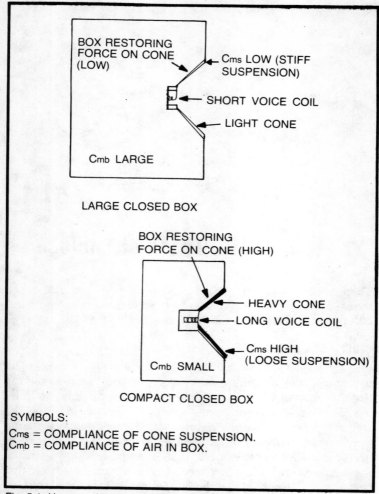

Fig. 5-1. How speaker type affects closed box size.

necessary restoring force on the cone, limiting its movement to a practical limit (Fig. 5-1). Even so, the engineers had to lengthen the voice coil so the cone could move farther than a conventional speaker without carrying the coil too far from the magnet. The engineers also found that by adding mass to the cone, they could lower its resonance to the subsonic band, if necessary, so that the speaker's resonance in the small closed box would occur near the bottom of the frequency range they wanted to reproduce. Although the added mass helped to smooth the frequency response of the speaker, it made the speaker extremely inefficient.

This inefficiency was the main disadvantage of the air suspension speakers. However, as the amplifier power race developed, audio power became cheap. The demands by stereo for two speakers, and later by quadraphonics for four, put a premium on compact speakers. These developments have increased the popularity of closed box bookshelf speakers over the last two decades.

The designers of the small closed box speakers credit their popularity to their quality, rather than their size. Air-suspension speaker manufacturers claimed that they attacked the problem of speaker design from a scientific approach and that the small box was a happy byproduct of their research. They said that the air behind the cone provided a more linear restoring force than the usual stiff suspension. Independent tests show that the best of these speakers have impressive specifications, low distortion, and a smooth frequency response. Closed box speakers are popular with hobbyists, too; the box design is relatively uncomplicated, and minor variations in speaker or box characteristics do not spoil the system.

Like all the other speaker systems, small closed box speakers have their disadvantages. The most frequently mentioned one is their power-hungry lack of efficiency, which means that you have to pay for a more powerful amplifier or receiver.

Another weakness, mentioned earlier, is that the special woofer used in compact boxes must have a longer voice coil and a greater cone mass than other speakers. The voice coil inductance and the heavy cone put such a strict upper limit on the woofer's frequency response that either great demands will be made on the tweeter, or else a 3-way system must be used. A 3-way obviously costs more than a simple woofer-tweeter system.

Another possible disadvantage of compact, closed box speakers is that in large rooms and at high sound levels, they have a more limited dynamic range than large efficient speakers. If you turn up the volume control for the soft part of a musical passage, your inefficient speaker may fluff the instantaneous peaks as it runs out of available amplifier power. Even if you have a superpowered amplifier, the same thing can happen; voice coil wire can stand only a certain amount of current before it overheats and gives up. Inefficient speakers must be operated at a higher voice coil current than efficient speakers to produce the same sound level. For most listeners, this is probably more a theoretical disadvantage than a real one.

Some of the advantages of early air suspension speakers have faded with time. Mechanical suspensions have been improved, and other kinds of box design have made great progress since the 1950's.

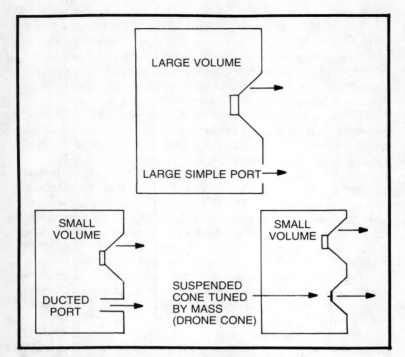

Fig. 5-2. Kinds of ported enclosures with action at resonance.

PORTED BOX ENCLOSURES

Like the closed box, the ported box, also called *bass reflex*, is simple to build; but, unlike the closed box, its action is complex. Except for the one difference that makes it another breed, it looks much like a closed box. That difference is a *port*, which is either a simple hole or a tube in one wall of the enclosure, usually the speaker panel (Fig. 5-2).

If you cut such a hole in a true infinite baffle, that hole would permit some of the sound to short circuit by taking a path that would produce cancellation at the frequency at which the length of that path was equal to one-half the wavelength of the sound. However, when you cut a hole in one wall of a closed box, the air in the box retains its ability to act as a spring, compressing and expanding in response to the piston action of the speaker. The air that is moving in and out of the port makes another "piston," one that vibrates in concert with the cone at some frequencies, and out of phase with others (Fig. 5-3).

This combination of a vibrating air piston and enclosed air in a box forms a resonator, properly called a *Helmholtz resonator* after the nineteenth century German physicist who first analyzed the behavior of tuned acoustical resonators. The frequency of the box resonance is controlled by the compliance of the air in the box, and by the mass of the air in the port. Like the speaker itself, the ported box is a tuned circuit in which a mass resonates with a compliance. When the tuned circuit of the speaker is closely coupled with the tuned circuit of the box by matching the resonance frequencies, the original

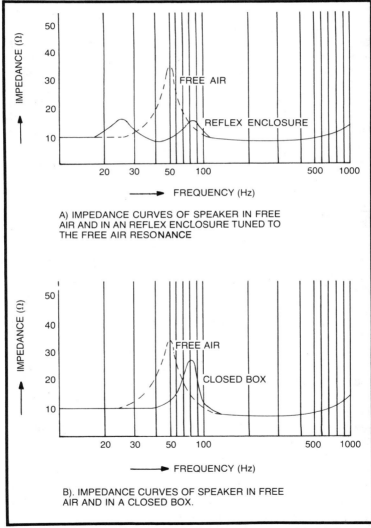

Fig. 5-3. Impedance curves of a speaker in a closed box and in a reflex.

resonance is replaced by two new resonances, one at a higher frequency than the original speaker resonance, the other at a lower frequency.

At the Helmholtz frequency of resonance, the air in the port vibrates easily, compressing and decompressing the air in the box with the cone. At this frequency, the output from the port is many times greater than the output from the cone because the action of the air in the port damps the cone so that the cone almost stands still. Because distortion varies directly with cone movement, this damping reduces distortion over the band of frequencies for which the port is effective, and it also reduces intermodulation and Doppler distortion at higher frequencies.

At the upper resonance, the air in the port vibrates in phase with the cone; but, because this frequency occurs well above the box resonance, the damping action is reduced. This is the frequency at which many reflex systems produce too much output, particularly small enclosures whose upper resonance occurs within the frequency range of the male speaking voice.

Below the box resonance frequency, there is a rapid phase shift in the port output; at the lower resonance the port radiation is 180 degrees out of phase with that of the cone. The out-of-phase radiation, plus the normal roll-off in response below speaker resonance, produces a cut-off that is sharper than that of closed box speakers.

The ported box can be tuned to the speaker by varying the size of the port. You could also change the tuning by varying the volume of the enclosure (Fig. 5-4). By adding solid material, such as glued-in wood blocks, you can reduce the volume and lower the compliance of the air in a box, but enlarging a finished box is usually impossible. You could change the operation of the air in the box from adiabatic to isothermal by adding fuzzy material for volumetric expansion. This is an inviting prospect, but the results can be unpredictable. The practical way to tune a ported box is to change the vent dimensions.

To increase the mass of the vibrating air in the port, you can change the port area. Inertance varies inversely with port size, so to lower the resonance frequency of a box, reduce the size of its port. The smaller port forces the air to move in and out with an increased velocity that causes the air on each side of the port to move with the port air, thus increasing the mass of moving air. There is a practical limit as to how far you can carry this box reduction. If the box is too small, then the required inertance in the port will be so great that the port must be a hole so small it will produce whistling noises as the air moves in and out. The practical lower limit of port area will vary with speaker size, tuning frequency, and power level.

When the audio world became aware that a duct or tube behind the port made it possible to tune small boxes to lower frequencies,

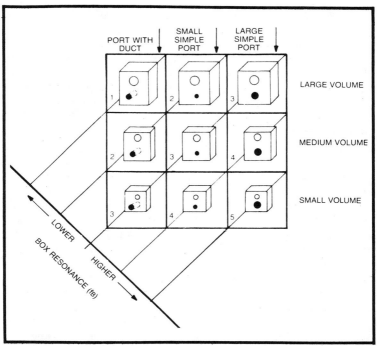

Fig. 5-4. How box volume and port affect frequency of box resonance. Diagonal lines and equal numbers indicate boxes with equal resonance.

this discovery was greeted with enthusiasm in the belief that it permitted smaller enclosures. Many statements in publications implied that the small box with a duct could give the same performance as a larger box with a simple port. Those comments confused tuning with total performance. They neglected the all important relationship of speaker compliance to box air volume compliance. In some cases, the small box can actually outperform a large box for a specific loudspeaker—but only if the box was designed to match the speaker's compliance and Q. In all cases, the design of small ported boxes is more critical than that of large boxes; minor changes in a small box can alter performance significantly. That is one reason that traditional advice to builders of reflex enclosures was "the bigger, the better." Now that the reflex is better understood, that advice has been changed.

One of the peculiarities of a reflex system is that port radiation does not vary with the size of the port. A small port will radiate just as much sound as a large port. When loudspeaker engineers recommend that the box be designed so that it can be tuned with a large port, it is not because the large port is more effective, but because a larger volume enclosure will usually perform better than a small one. Also, a port that is too small can produce the unmusical noises mentioned earlier by its high air velocity. In some cases, an extremely small port can impede the reverse air flow, producing more distortion than a closed box would produce.

The small port is the Achilles heel of the ultracompact vented enclosure, but this weakness can be solved by substituting an extra diaphragm for the air in the port. The port diaphragm is made like a speaker without a magnet or voice coil assembly. It can be tuned by adding or subtracting mass from the vibrating diaphragm, just as the ported box can be tuned by adding or subtracting from the mass of the vibrating air. H.F. Olson described the first "drone cone" in 1955. The "drone cone" has since appeared under such names as *passive radiator* and *auxiliary bass radiator*. The passive radiator's main advantage is that it can be used to tune a small box to a low frequency. The passive radiator also blocks reflected mid-range sound that could pass through an open port if it were not absorbed inside the box.

The ported system's most important advantage is its damping control on the speaker. The ported system also offers two to three times the efficiency of a closed box speaker of the same size, and low frequency cut-off by permitting a more efficient woofer. Or, by altering the design criteria, a ported box can have the same efficiency and cubic volume of a closed box, but then its bass range will extend about a half octave below that of the other. A woofer designed for a ported system does not have to have as heavy a cone or as long a voice coil as the sealed box woofer, so its response can extend into the mid-range, making a simple woofer-tweeter system a practical possibility.

The biggest disadvantage of the bass reflex is that it requires careful design to permit it to achieve its theoretical advantages. In fact, many ported box speakers have limited low bass response and nasty resonant peaks that have earned them the name of "boom boxes." A manufacturer must either institute careful quality control on the production of his bass drivers, or else compute the box dimensions and tuning for each individual speaker to insure consistent performance from ported box speakers.

A less obvious disadvantage of the ported box is its tendency to unload the speaker below resonance. At extremely low frequencies, turntable rumble or other faulty components can produce subsonic pulses that overload the speaker, causing distortion, or even damage. The solution to this problem is either to use good equipment with a ported box system, or to have a low-cut filter to remove unwanted low frequency disturbances.

LABYRINTHS AND TRANSMISSION LINES

A labyrinth is a tuned pipe with the driver at one end and the other end open (Fig. 5-5). When the wave from the driver reaches the end of the pipe, it expands into the room, causing a sudden pressure drop. This drop in pressure is reflected back through the pipe to the speaker. At the frequency where the length of the pipe is equal to a quarter wavelength ($\lambda/4$) of the sound, the air at the mouth of the pipe is at minimum velocity, but maximum pressure. At that frequency, the change of pressure is also at a maximum, and the reflection to the speaker of the rarefaction produces maximum damping. So a labyrinth that is a quarter wavelength at the speaker's resonance frequency will damp the speaker at resonance, an action similar to that of the bass reflex.

At twice the frequency of resonance, the frequency that corresponds to a half wavelength, the same length of pipe produces high velocity but low pressure at the mouth. This situation produces little or no anti-resonant action, permitting the speaker to vibrate freely. Phase reversal at the half wavelength adds to the cone output, which can cause a response peak at that frequency. By carefully matching the pipe to the speaker, the boost in response can be used to fill a valley in the speaker's normal response curve.

The first labyrinth was developed by Benjamin Olney for Stromberg-Carlson in 1936. Olney found that when he put a speaker with a 50 Hz resonance in the labyrinth, the frequency of resonance dropped to 40 Hz. He used a pipe that damped the speaker's resonance and added to the output at 75 to 80 Hz, the half-wave frequency. This half-wave frequency was low enough that the boost in output helped bass response without creating an annoying boom.

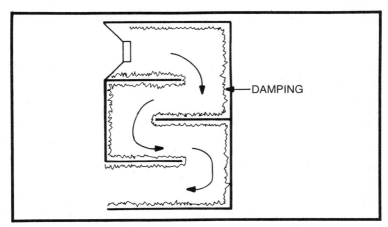

Fig. 5-5. Labyrinth.

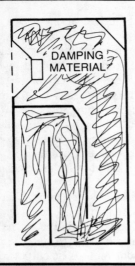

Fig. 5-6. Transmission line.

Olney's next problem was a series of peaks farther up the frequency scale. These peaks occurred as harmonics of the fundamental pipe resonance. He lined the pipe with absorbent material that, combined with bends in the pipe, cut off the response at the pipe mouth above 150 Hz.

Although the labyrinth has always had a following among experimenters, it was not a commercial success. It had to complete in size and cost with enclosures that were simpler to build, such as the bass reflex and the closed box. Stromberg-Carlson re-introduced labyrinths to the component hi-fi market in the 1950's, but later they abandoned consumer products, and the labyrinth disappeared again. Then, in the mid-1960's, it reappeared as the "transmission line."

A transmission line is a stuffed labyrinth (Fig. 5-6). The theory behind the transmission line was developed by Prof. A. R. Bailey of the Bradford Institute of Technology, Bradford, England. Bailey suggested that bass-reflex enclosures, with their sharp cut-offs, were likely to produce ringing, similar to electronic devices with sharp cut-offs. Bailey reasoned that an infinitely long acoustic line behind the speaker would absorb the backwave without causing troublesome reflections that could produce standing waves behind the cone, so he substituted stuffing for infinite length. He used long-fiber wool, to reduce the pressure changes that occur in a labyrinth. Extremely low frequency waves were not absorbed by the wool but emerged from the mouth of the line to augment the speaker's low bass response.

Bailey's design has been adopted and altered by various speaker manufacturers, including IMF, Infinity Systems, ESS, Radford, and others. Some of these designs use the output from the end of the line for bass reinforcement. Others use the line as an absorptive device only.

Labyrinths usually had a cross-sectional tube area that was equal to or greater than the speaker cone area. Transmission lines are often tapered, with the large end having a greater area than the cone, and the small end having an area that is smaller than the cone.

Robert Fris, an English experimenter, has introduced a variation of the transmission line which he calls a "Daline," an acronym for "*d*ecoupled *a*nti-resonant *line*" speaker. Fris decouples the driver from the line by adding a cavity directly behind the driver (Fig. 5-7). The volume of this cavity is made right to tune the cavity resonance up to a half octave above the line resonance. At frequencies above the pipe's anti-resonance, the cavity absorbs the rear radiation of the speaker. Below that frequency, the pipe takes over and loads the speaker. Fris claims that his line has no problem with harmonic resonances in the pipe because these frequencies are absorbed before they get to the pipe. He uses 4 in. to 6 in. speakers to drive the Dalines, saying that in a Daline they can equal the performance of much larger speakers and have better transient response.

A smooth impedance curve is characteristic of a well designed transmission line, as contrasted to the single peak in the impedance curve of the closed box, or the double peak in the impedance curve of a ported box. The value of a smooth impedance curve is open to debate among speaker engineers. Those who favor closed boxes or ported systems say that the shape of the impedance curve makes very little difference in the output of the speaker because modern amplifiers operate as constant-voltage sources. Only a constant current device would produce a frequency response that would follow the speaker's impedance curve, they say. Transmission line advocates reply that solid state amplifiers can not deliver their rated

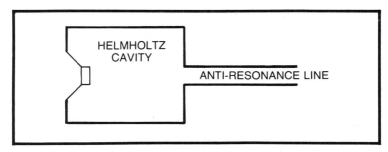

Fig. 5-7. Fris Daline enclosure.

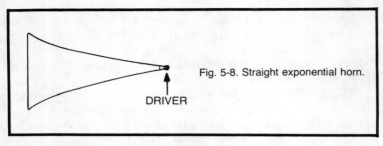

Fig. 5-8. Straight exponential horn.

power into a high-impedance load, to which the other team argues that impedance peaks occur at points of resonance and it does not matter if you do not deliver full power into a resonance because the added efficiency at resonance takes care of the reduced power. And so on.

The disadvantages of transmission lines are obvious. They are large, they require a complicated structure, and they can be unpredictable. Fine tuning a transmission line is largely a matter of trial and error, but for some audio fans, this experimenting is part of the challenge and the fun.

HORNS

A horn acts as an acoustical transformer. It matches the high acoustical impedance at the driver to the low impedance of the room air by its smooth rate of increased cross-sectional area from the driver cone to the horn mouth (Fig. 5-8). Just as mismatching electrical impedances can cause inefficient power transfer, direct radiator speakers lose acoustical efficiency because of the sudden transition from high pressure at the cone to low pressure in the room.

An acoustical megaphone, the kind used by cheerleaders, has some of the virtues of a horn. However, being straight sided, it does not permit the waves to expand at a constant rate; as the area of the megaphone increases, the distance between points where the area doubles also increases. A true horn has a flare which forces the sound waves to expand at a constant rate. Because of its impedance matching characteristics, a horn offers much higher efficiency than other types of speaker enclosures.

In the early days of hi-fi, many devices that were *called* horns were offered for sale. Some of them did not operate as true horns, the usual deficiency being size. They were too small to act as a horn in the low bass range. Some had obnoxious resonances or distortion because of poor design.

The size required for a well designed horn for a bass driver is tremendous: some horns are thirty feet in length. The greatest commercial use of horn-loaded bass drivers today is in theater sound systems.

One horn, developed by Paul Klipsch in 1940, is still in commercial production. Klipsch reduced the size of a bass horn by folding it and using the corner of the room as an extension of the horn. His design, called the "Kilpschorn," is an exponential horn with a 40 Hz cut-off.

The Klipschorn was followed by numerous pseudo-horns that consisted mainly of folded baffles such as that of Fig. 5-9. This kind of enclosure was popular in the early days of hi-fi, but some of them performed poorly, with little resemblance to the sound of a true horn.

The advantages of the horn are its efficiency and, as a by-product of efficiency, its dynamic range. The horn also permits lower distortion at higher output than any other enclosure because of its high damping on the driver.

Some disadvantages are obvious, such as its complexity and cost. The possibility of phase distortion is another, more controversial, shortcoming of horns. Phase distortion occurs when the sound from one driver lags behind or leads that of another driver. Phase distortion in horns was first noticed by Hollywood sound men while

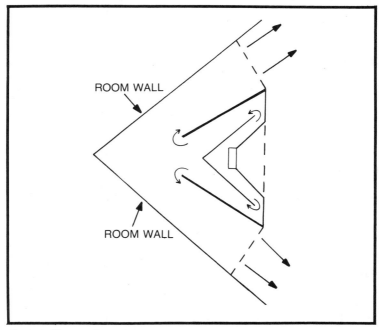

Fig. 5-9. Direct radiator speaker in folded "horn," 1950's.

they were listening to a tap dancing sequence by a film star. The sound of the tap dancing contained an echo after each tap. They traced the source of the echo to the speaker, a large two-way horn. The bass horn was eight feet longer than the treble horn. This difference can be compensated for in a large theater system by moving the treble horn back in space to equalize the path length, but the problem is more serious for folded horns designed for home use.

These disadvantages make horn construction for the amateur a discouragingly difficult project. Just as most commercial attempts at producing a folded horn for home use have failed, horns produced by amateurs have almost invariably ended their lives in more useful roles, such as dog houses.

HYBRIDS

In addition to the four enclosure types described here, you may see an occasional variant. The open back enclosure, popular with manufacturers of old console radios, is used only on low-fi equipment today, although one manufacturer of high fidelity speakers used to make a stuffed open back enclosure. The flat baffle is used in full range electrostatic models and, with a few dynamic speakers, permits the back wave to be reflected by room walls and other objects. And you will sometimes see bass reflex horn hybrids. That's about it.

WHICH ENCLOSURE IS BEST?

Even if one of the enclosures described here could be proven to give the best performance in every system, it would not automatically become the universal choice. The bass horn is a good example of a high performance enclosure that is made rare by its cost and complexity.

In choosing an enclosure, you should consider your living quarters, your preference in sound, and even your personality. Ask yourself this: if you had some money to invest, would you put it into a venture such as a search for an oil well that could make you rich—or bankrupt you? Or would you put it into a government insured savings account? The answer to that can tell you something about the kind of speaker enclosure you should build.

Popularity may be a poor criterion of what is good, but it does show which enclosures are more practical for most people, or at least for manufacturers. The table in Fig. 5-10 summarizes the chief advantages of the two most commonly used boxes for speakers, closed boxes and reflexes. You can look over this summary and get a rough picture of the strong points of each kind, but only a rough one.

ADVANTAGES OF CLOSED BOX	ADVANTAGES OF REFLEX
1. SIMPLE TO DESIGN AND BUILD. 2. MORE GRADUAL CUT-OFF RATE OF 12 dB/OCTAVE. 3. GOOD FOR SUB-MINIATURE ENCLOSURES. 4. CAN USE WITH HIGHER POWERED AMPLIFIERS BECAUSE IT DOESN'T UNLOAD SPEAKER AT LOWEST FREQUENCIES.* 5. SPEAKER VARIABILITY HAS LESS EFFECT ON PERFORMANCE.	1. HIGHER OUTPUT AND LOWER DISTORTION IN OCTAVE ABOVE fB. 2. GREATER BASS RANGE OR HIGHER EFFICIENCY FROM EQUAL BOX VOLUME. 3. CAN USE SINGLE CONE SPEAKER WITH REDUCED PHASE PROBLEMS. 4. LOWER COST BECAUSE OF # 3. 5. GREATER CHALLENGE—GREATER REWARD IF DONE RIGHT.

* THIS ADVANTAGE FOR THE CLOSED BOX CAN BE NEUTRALIZED IF AN EQUALIZER IS USED WITH REFLEX SYSTEMS.

Fig. 5-10. Summary: Closed Box vs. Reflex.

Some readers may question whether Point 5 for the reflex, the greater challenge, isn't really a disadvantage. It can be. For the management of a company trying to make a profit, a challenge is the last thing needed. The challenge can be translated into higher costs because of closer quality control of speakers or tuning problems. Here is one advantage the owner/builder has over a speaker company. Tinkering with a ported box can be fun for the one, and nothing but expense for the other. Anyone who likes to experiment with the unfamiliar can build a transmission line and have a speaker system that may not be better than the typical speaker-in-a-box, but will certainly be different.

Those who treasure classic simplicity and dependability above mystery and theoretical gains will appreciate the closed box. You still have to know which way to aim the design, but the target is bigger. Anyone who goes through projects at a gallop should choose the closed box.

One disadvantage of the reflex, suggested in Fig. 5-10 by Advantage 4 for the closed box, should be considered even though it need not be a disadvantage. The speaker in a reflex enclosure is highly damped at resonance and for a band of frequencies above resonance by the air piston in the port working in phase with the cone. The two pistons are compressing the air in the enclosure at the same time, and the high Q port action damps the cone. Below resonance, there is a rapid phase shift so that at the lower impedance peak, the air in the enclosure appears to be in series with the cone. As the cone moves in, the air in the port moves out and vice versa. The cone is unloaded at extremely low frequencies where great cone movement is required.

In a properly designed system this out-of-phase action should be low enough in frequency that the signal could be chopped off in the amplifier, producing a sharp roll-off in subsonic bass without affecting music frequencies. It would be a good thing if all amplifiers had this kind of cut-off, but many do not. The practical effect of this situation is that reflex speakers can be driven into distortion or damaged by powerful amplifiers with extended bass response. In some cases, it is not even the fault of auxiliary equipment; the amplifier is unstable at low frequencies. The best cure for this potential problem is a reflex equalizer, which is an outboard filter that can be easily added to the amplifier. Anyone who owns a super amplifier and likes his music loud should consider adding the equalizer. One is described in Chapter 7, and a schematic diagram is shown in Chapter 16.

Your choice of enclosure will be partly determined by the kind of speakers you buy or own. If you have not decided yet, try to choose the kind of system that fits your needs and your temperament. Your choice should be governed by different reasons than those of a manufacturer, but the choice you make will not necessarily be different. After all, the decision as to which enclosure is best must ultimately be made by you and by you alone.

6

Closed Box Systems

The closed box is the simplest speaker enclosure you can make, both in principle and in construction. In the audio world there is apparent agreement among engineers as to which priorities should be observed in closed box design. That alone is a good indication that this breed of enclosure is pretty well understood.

Its simplicity makes the closed box a happy choice for the novice or anyone who has little in the way of equipment. Minor variations in design will not spoil the sound. You do not face a set of design questions about ports, tuning frequency, pipe length, or other complicated variables; there is just one main question—"How big?" When you buy a woofer, words such as "big" or "small" are ambiguous. You may wonder "How big is big?" or, more to the point, "How big is just right?"

In free air, a speaker's resonance is determined by the mass of the moving parts which vibrate on the compliance of the suspension. A simplified electrical analogy of this is an inductance in series with a capacitance, as shown in Fig. 6-1A. R is the resistance of the air load. If the value of M or C_{ms} is increased, the resonance frequency is lowered. If the value of C_{ms} is reduced by making the cone stiffer, the resonance will be higher in frequency, just as it would be in an electrical circuit.

When you put a speaker into a closed box, the trapped air in the box pushes against the cone when the cone moves inward, and creates a partial vacuum when the cone moves out, causing the outside air to push against it on that half cycle. The cone is no longer as free to vibrate as it was out of the box; in effect, it is stiffer in the

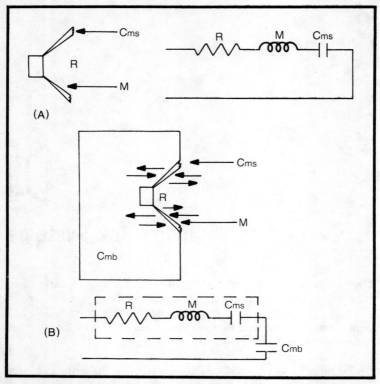

Fig. 6-1. Equivalent acoustical circuits of a speaker in free air (A) and in a closed box (B).

box. This stiffness raises the frequency of resonance. The system resonance of a speaker in a closed box is always somewhat higher than the free air resonance of the same speaker.

Going back to the simplified electrical analogy, the air in the box acts like another capacitance. When the speaker goes into the box, it sees the air in the box as a capacitance in series with its own compliance capacitance (Fig. 6-1B). Two capacitances in series always produce a net capacitance that is lower in value than either of the single capacitances. Just as a reduced electrical capacitance shifts a resonance upward in frequency, the analogous capacitance of the air in the box raises the frequency of resonance for the speaker.

Because the degree of air stiffness depends on the cubic volume of the enclosed air, the size of the box controls the final system resonance of a speaker system. Below resonance, the output will drop at a steady 12 dB per octave, so the point on the frequency band where resonance occurs is of considerable importance. A certain volume of air has no special value for stiffness by itself. The determining factor is the size of the piston coupled to it. The air in a box will resist the movement of a large piston more than it will resist a small one. In fact, the compliance of the box air varies inversely with the square of the driver's cone area. Consider how a box would affect the compliance of a typical 10 in. speaker compared to a 5 in. speaker. The effective cone area of a 10 in. speaker is about 4 times that of a 5 in. speaker, so the air in a box of a certain size would act against the cone of a 10 in. speaker with 16 times the pressure that it would act against the 5 in. speaker's cone. The formula for box compliance is:

$$C_{mb} = \frac{V}{dc^2A^2} \quad (cm/dyne)$$

Where:
    $V =$ the cubic volume of the box
    $d =$ the density of air
    $c =$ the speed of sound in air
    $A =$ the effective area of the driver

If you are ever tempted to squeeze a large woofer, say a 15 in. size, into a compact box, remember this formula: it proves that you can not get a low frequency system resonance that way (Fig. 6-2). I once tested three speakers that had identical free air resonances of 50 Hz in boxes of several sizes, the rated diameters of the speakers being 4 in., 6 in., and 8 in. In a 3 ft.³ box, the resonances were fairly close, ranging from 52 Hz for the 4 in. speaker to 60 Hz for the 8 in., but in smaller boxes the effect of the box on the speakers began to show. In a 0.5 ft.³ box the resonances were 59 Hz for the 4 in.

Fig. 6-2. The formulas are right: you cannot put a big speaker in a little box.

speaker, 85 Hz for the 6 in. speaker, and 99 Hz for the 8 in. speaker. So while the stiffness of the air in a 0.5 ft.^{3} box against the 4 in. speaker is only enough to raise the frequency of resonance by about 20%, it is great enough with the 8 in. speaker to raise the frequency of resonance 100%. How much a particular box size will raise the frequency of resonance depends on the ratio of box air compliance (C_{mb}) to cone compliance (C_{ms}).

The term "V_{as}," when applied to a speaker, indicates the volume of air that has a compliance equal to the compliance of that particular speaker. When a speaker is installed in a box that is equal in volume to the speaker's V_{as}, the situation is that of an electrical circuit with two equal capacitances in series. The formula for capacitances in series is:

$$1/C = 1/C_1 + 1/C_2$$

When $C_1 = C_2$, as in the case mentioned above, the resulting capacitance is equal to one half the value of each. In the formula for resonance:

$$f_s = \frac{1}{2\pi\sqrt{M\,C_{ms}}}$$

If all values remain the same except C_{ms}, and that is halved, the new resonance is equal to 1/0.707, or 1.414 times the original resonance. So if your speaker has a free air resonance of 35 Hz, and you put it into a box with a volume equal to V_{as}, the system resonance will be 1.414 × 35, or about 50 Hz. Whether or not it would be desirable to put the speaker into such a box would depend as much on the speaker's Q as on its frequency of resonance.

Figure 6-3 shows how the Q of a speaker in a closed box determines the shape of the response curve near resonance. If the Q is very much greater than 1, there is a nasty peak in the output curve that produces a boom, or one note bass with hangover, after the signal has ended. On the other hand, with a Q of 0.5, as shown, the bass response starts cutting off at about two octaves above resonance and is down 6 dB at resonance. Most closed box engineers aim for a Q of 1, which gives a slight bump in response above resonance, with the curve passing through the 0 dB line at resonance. This gives the most extended bass response without significant ringing. The 0 dB line on the chart in Fig. 6-3 represents the speaker's average output in the midfrequency band.

Some designers prefer a Q of 0.707, the condition that gives the flattest curve, with no peaking, and slightly better transient response than a speaker with a Q of 1. A speaker with a 0.707 Q can take bass boost without problems. This boosting can restore the low frequency bass to a flat response, but the boost circuit should be

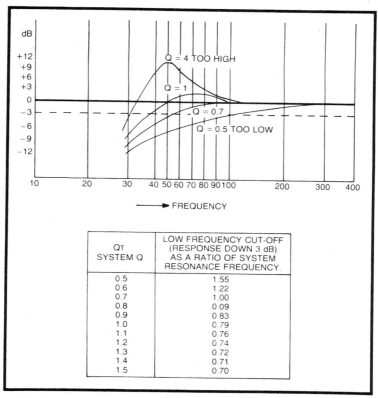

Fig. 6-3. Relation of System Q to low frequency response and cut-off. To find the cut-off frequency, get the ratio from the chart that is correct for system Q. Multiply that ratio by the system frequency, f_c.
Example: System has Q_T of 0.9, f_c = 70 Hz.
   Cut-off f = 0.83 × 70 Hz = 58 Hz

tailroed to the speaker's response curve to avoid producign a peak at middle bass frequencies. A Q of 1 for the typical closed box speaker is more practical for most speakers.

Regardless of the theoretical implications, there are some speaker systems in which a relatively high Q can be tolerated by the ear. For example, if the system resonance is very low, say below 50 Hz, most listeners will not be nearly as offended by a slight boominess as they would be were the boom at 100 Hz or higher. Although a low resonance and a low Q usually go together, some speakers have a free air resonance below 20 Hz and a Q of 0.5 or higher. When installed in a small box to produce a system of resonance of about 50 Hz, these speakers will have a final Q that is greater than 1, but it is not objectionable because the frequency is so low that the ear is less

RATED DIAMETER OF SPEAKER (IN.)	RANGE OF VOLUMES FOR CLOSED BOX (FT.^{3})
4	0.1 – 0.25
5	0.15 – 0.3
4 × 6	0.15 – 0.3
6	0.25 – 0.75
5 × 7	0.25 – 0.75
8	0.5 – 1.5
6 × 9	0.5 – 1.5
10	1.0 – 3.0
12	2.0 – 5.0
15	5.0 – 10.0

Fig. 6-4. Range of required box volumes for speakers of various diameters.

sensitive to peaks here. However, a speaker with a high resonance can sound boomy with a Q of about 1. What looks right or wrong on paper may not work out exactly that way in real speaker systems.

If you measure the Q of your speaker before and after you put it into a box, you will find that the box raises the Q by about the same degree that it raises the frequency of resonance. This implies that the manufacturer should adjust the Q by choice of magnet, and adjust the free air resonance by controlling the mass and compliance of the cone so that they work well together. If the manufacturer does this, then in a box of suitable size, the speaker will have a desirable frequency of resonance and Q_t (total system Q). This also suggests that some speakers may be impractical for use in a closed box. That is right, but some kind of a compromise solution can be worked out for most high compliance drivers.

HOW TO FIT BOX VOLUME TO SPEAKER

If you already have a woofer, the problem is to find the right size box for it. Sometimes the procedure should be reversed; the problem is how to find the right speaker for a given box. This second approach would be in order if there is to be a strict limit on the size of the cabinet that can be accommodated.

To get a rough idea of what box volumes are appropriate for speakers of various sizes, look at Fig. 6-4. These figures for box volume show a considerable range of values because of the normal variation in compliance, even though this table is based on speakers that are labeled "high compliance."

If you have no test equipment, you can use Fig. 6-5 or 6-6 to get a rough estimate of the system resonance of a 4 in. to 12 in. speaker in a certain size box. The lines and shaded areas on the simplified design charts are based on tests on several high compliance speakers of each size listed. The testing results were averaged for each size, and system resonance calculations made for the "average" speaker of that size. For example, six different 10 in. woofers were tested. The average frequency of resonance for these speakers was 35 Hz, and the average V_{as} value was 4.5 ft.^{3} A box design worked out for such a speaker would not fit any one speaker perfectly, but should give fairly good performance with all but one or two of the six. You may note that the shaded area on the chart for 4 in. to 6 in. high compliance speakers is much narrower than the shaded area for larger woofers: there was much less variation in the characteristics of the smaller speakers. You can use the simplified design chart for

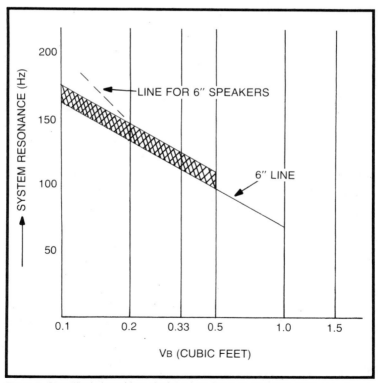

Fig. 6-5. Simplified closed box design chart for 4, 5, and 6 in. high compliance speakers where free air resonance is unknown.

147

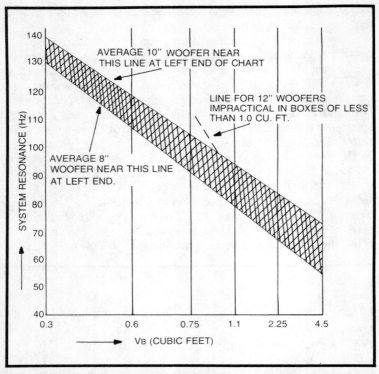

Fig. 6-6. Simplified closed box design chart for 8, 10, and 12 in. high compliance woofers where free air resonance is unknown.

the small speakers with more confidence that the system resonance will be similar to that predicted. The typical 4 in. speaker tested had a free air resonance of 105 Hz and a V_{as} of 0.15 ft.^{3} For 5 in. speakers, the figures were 95 Hz and 0.2 ft.^{3} For speakers with roll suspensions, these values were fairly consistent, except for one or two variations.

But what is the best system resonance for a certain size speaker? For the typical speaker with a fairly large magnet and a high compliance, the system resonance must be some frequency well above the free air resonance to give the proper Q_t value and to add the necessary degree of stiffness to a loosely suspended cone.

If you have no test equipment, you can follow this rule of thumb for the smallest speakers: set the system resonance at 1.414 times the free air resonance. To do this, you would make the box air compliance equal to the compliance of the speaker's suspension, and make the box volume equal to V_{as}. According to the data on the speakers tested, that would be a box of 0.15 ft.^{3} for a 4 in. speaker, and 0.2 ft.³ for a 5 in. speaker. These combinations would give system resonances of almost 150 Hz and 135 Hz respectively. If these values seem high, that's too bad; such are the facts of life, except for the rare small speaker that can go lower in frequency.

With the larger speakers, there is more variation, but the chart in Fig. 6-6 will give some indication of what to expect.

HOW TO USE TEST DATA TO GET THE RIGHT BOX SIZE

Measure the values of f_s, Q, and V_{as} for your woofer (Tests 1, 3, and 6, Chapter 3). Find the optimum Q_t/Q ratio from 1 divided by the Q of your speaker (1/Q). Or simply look up the figure for your speaker's Q on the left margin of Fig. 6-8. For example, if the Q of your speaker is 0.5, the optimum f_c/f_s ratio will be 2 (on right margin). The horizontal line for a speaker with a Q of 0.5 crosses the diagonal line above 0.33 at the base for the optimum V_B/V_{as} ratio. Multiply this figure by the V_{as} of your speaker to get the box volume that will give a Q of approximately 1. In this example, if V_{as} were 6 ft.³, the box volume would be 2 ft.³ You can use the f_c/f_s ratio to predict the system resonance of the speaker in that size box. For the speaker with a free air resonance of 25 Hz that is mentioned above, the system resonance in the 2 ft.³ box would be about 50 Hz.

Sometimes the answers are not as clear cut as in the hypothetical example given here. Here is the test data on a 12 in. high compliance woofer with a foam roll surround and a 54 oz. magnet:

$$f_s = 24 \text{ Hz}$$
$$V_{as} = 12 \text{ ft.}^3$$
$$Q = 0.31$$

If we make a Q_t of 1 as the first condition, we must look for a box volume that will raise the Q by at least a factor of 3, but that means that the system resonance would be raised by the same degree, to about 75 Hz. This in turn means that the bass response would roll off at 12 dB per octave below that frequency, which seems a waste of low frequency ability in this 12 in. woofer.

We can choose a larger box to keep f_c low. It would be desirable to keep f_c down to 40 Hz, which pretty well covers the musical range, but then Q_t would be about 0.5 (40/24 = 0.5/0.31). The speaker would be overdamped and the bass response would start to roll off a couple of octaves up the scale. This dilemma tells us that there is no guarantee of good bass response in a closed box just because a speaker has low free air resonance and looks expensive. The specifications of this driver are more typical of speakers designed for use in ported boxes, and that was surely what the manufacturer had in mind for it.

But suppose you have such a speaker. What do you do? The best solution would be to use it in a ported box which needs a low Q speaker; a reflex permits a greater bass range than a closed box of equal volume. However, this speaker would probably work satisfactorily in a closed box system with a Q of about 0.7. This would raise the resonance to about 54 Hz, a good figure except that with a Q of 0.7, the response will be down 3 dB at resonance. Let's assume that you have an amplifier with a multiband frequency control so that you can correct for the drooping low bass response without boosting the middle bass. In that case, this would be a good choice, so let's follow through on it.

If Q_t is to be 0.7, then Q_t/Q will be 0.7/0.31, or 2.26. To use the design chart, ignore the Q figures on the left margin and look for a 2.26 ratio on the right. Note that 2.25, which is close enough to 2.26, gives a V_B/V_{as} ratio of 0.25. For a Q_t of 0.7 then, this speaker should have a box volume of 3 ft.^{3} (12 ft.^{3} × 0.25 = 3.0 ft.^{3}).

Before you would go ahead with a 3 ft.^{3} box for such a speaker, you should consider one additional factor: if you add stuffing to the box, the air inside the box will change its behavior from adiabatic (constant heat) to isothermal (constant temperature). This means that the velocity of sound is reduced and the wavelengths shortened, making the box appear bigger than its dimensions indicate. You can probably add 20% to 25% volume in this manner, so you can make the box slightly smaller than the suggested 3 ft.^{3} You would undoubtedly be safe in lowering the volume by at least 16%, to 2.5 ft.^{3} or less, and then stuffing the box with Dacron or fiberglass. Be careful not to overstuff, or you may get back to a Q that is too low for a full bass response.

PROJECT 3: A 3-WAY CLOSED BOX SPEAKER SYSTEM

For this project, we will try to select a system that will give good response in a cabinet no larger than 2 ft.^{3} (Fig. 6-7). Looking at Fig. 6-4, it appears that a 10 in. woofer is the largest one we should use. To work well in a 2 ft.^{3} box, any larger woofer would have to have an excessively heavy cone and severe inefficiency, or it would cut off at a higher than desirable frequency.

We could match this box to many woofers, but the first one we will try is a Radio Shack model 40-1331-A. Here are the test data on that woofer:

$$f_s = 40 \text{ Hz}$$
$$V_{as} = 5 \text{ ft.}^3$$
$$Q = 0.8$$

The relatively high Q is no surprise because of the magnet size (10 oz.) for this size speaker. Actually, it will be all right because we will not have to move too far with f_c. But will it work in a 2 ft.^{3} box?

Fig. 6-7. Front view of Project 3.

The free air resonance turns out to be higher than expected, which suggests that this speaker is abnormal in that respect. This is no great problem though; this particular specimen probably has a slightly stiffer cone than the average speaker of this mode, which would mean a higher figure for both Q and f_s. But if that assumption is correct, any speaker of the same model should work equally well in a closed box designed for this one. If another specimen has a looser suspension and a lower f_s, it will also have a lower Q, but a larger V_{as}, so the quantities will change in a complementary manner that will still require the same box design as long as there are no changes in magnet size, cone mass, etc.

Since we are designing for a 2 ft.^{3} box rather than for a box of optimum volume, we look for data on a box that has a specific V_B/V_{as} ratio instead of a certain Q ratio. Here we want volume ratio of 2/5 or 0.4. Figure 6-8 shows that the box will raise the Q between 1.75 and 2 times the free air value. That will make Q_t of about 1.5, which would be too high.

Here we have the opposite situation from that shown earlier for the speaker with a 54 oz. magnet. If the magnet here is smaller than we would like, that's a fact of life and we must either use this one, or be prepared to pay more for the driver. In this case we will use it;

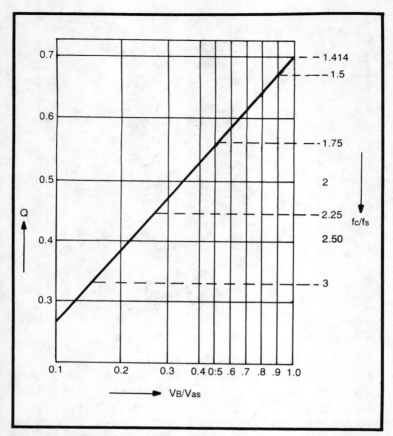

Fig. 6-8. Closed box design chart.

after all, most speaker systems are made up of this kind of compromise. We will stuff the box with damping material and, if necessary, use a damping pad over the woofer to control the Q. The figures indicate that by doing that we can keep f_c down in the 65 Hz range, well below the dangerous 100 Hz frequency which is the region of male speech, and getting the Q down to 1 is not mandatory if the frequency of resonance is low enough.

For the mid-range we will demand a speaker with a low Q because a peak at a few hundred Hz is much more annoying. A full range 4 in. speaker, Radio Shack 40–1197, gave these test values:

$$f_s = 95 \text{ Hz}$$
$$V_{as} = 0.17 \text{ ft.}^3$$
$$Q = 0.32$$

To keep Q_t at 0.7 or lower to avoid peaking, we will want the Q_t/Q ratio to be no greater than 2.28. From Fig. 6-8 we find V_B/V_{as} ratio to be 0.25 for a Q and resonance ratio of about 2.25. So V_B should be 0.0425 ft.^{3} or greater.

Looking around for a suitable ready-made enclosure for the mid-range speaker, we find that with a little rasping of the inside edge, a 3 in. I.D. tube which is available from carpet dealers, will accept the speaker. The internal depth of the cabinet (10⅜ in.) allows no more than that, plus the thickness of the front panel (¾ in.) for tube length, but we will try 9 in. to leave some clearance at the back. The volume of such a tube is about 0.037 ft.^{3}, a little short of the ideal; but, by loosely filling the tube, we can gain at least 20% volume, bringing our effective V_B up to about 0.044. That should be close enough, because with front mounting, a 4 in. speaker occupies an insignificant space in an enclosure.

For the tweeter, we will choose a closed back driver that needs no enclosure. We find that the CTS tweeter has a resonance at 1000 Hz, so if we cross over at 5000 Hz, this presents no problem.

This concludes the speaker selection and enclosure design. Construction details appear in Chapter 15.

PROJECT 4: A 2-WAY CLOSED BOX SPEAKER SYSTEM

The woofer used in this project (Fig. 6-9) has a free suspension with an unusually low resonance for a 10 in. speaker. Because of that free suspension, you should inspect the flatness of the spider if the speaker has been stored for any length of time. This precaution holds for most of the high compliance speakers listed in this book, particularly those with treated cloth suspensions. If the speaker is stored on its back or face for a long time, the cone can drift out of position. The cure is to reverse the speaker's position and allow the cone to drift back again before installing the speaker in the enclosure. Once the cone is in a vertical position, it will hold its proper position indefinitely, but a drifting cone will move the voice coil away from the region of maximum magnetic field, which decreases the damping on the cone and increases distortion.

You can identify a problem of cone drift either by the position of the spider, or by a change in Q as measured by Test 3. The position that gives the lowest Q is the right one.

Tests on the woofer showed these values:

$$f_s = 24 \text{ Hz}$$
$$Q = 0.45$$
$$V_{as} = 8 \text{ ft.}^3$$

Going to the design chart, it appears that the ideal box volume would have a volume of about 2 ft.^{3}, which would give a system

Fig. 6-9. Project 4 completed.

resonance 2.25 times 24 Hz, or 54 Hz. This assumes no damping material in the box, so we will make the box volume **1.5 ft.**^{3} Even if the Q is slightly more than 1, the hangover will not be particularly noticeable at this range of frequencies. Looking at the design chart for a V_B/V_{as} of 0.19, which would be equivalent to a **1.5 ft.**^{3} box, the system resonance will be 2.45 × 24 Hz, or 59 Hz and the Q about 1.1. However, after lining the walls with 2 to 3 in. of fiberglass, the actual measurement of the system resonance was 55 Hz. This indicates a Q of about 1, generally considered ideal for the most extended bass response without significant hangover. Going to the chart in Fig. 6-3, the bass cut-off appears to be about 0.79 × 55 Hz, or 44 Hz.

This system should be rated at 4Ω impedance because a woofer of lower than normal impedance is coupled to two tweeters wired in parallel. The woofer could be used with just about any 8Ω tweeter because the imbalance in impedance would be on the side to favor the woofer. Even if used with an 8Ω tweeter, the woofer would make a system that should be rated at 4 Ω to prevent anyone from using two systems in parallel and producing a low impedance load to solid state amplifiers.

The tweeters are larger than most tweeters, so they are stacked vertically to aid dispersion. A typical problem with 2-way systems becomes apparent when you start matching a tweeter to a woofer that is 10 in. in diameter or larger. To keep the crossover point low enough, you need a large tweeter, but to get the best dispersion at the top, you need a small one. Some small diameter tweeters are satisfactory for a 1000-1200 Hz crossover.

The crossover network is designed as a 1200 Hz dividing circuit between the single woofer and the two tweeters. The calculated value of the capacitor for a 4 Ω tweeter system was about 33 μf, but tests indicated that response was adequate with a 12 μF capacitor. Most speakers are overly efficient in the 1000 to 2000 Hz band, so if the tweeter's output is cut a bit in that band, it should cause no holes in the response curve. The woofer has good response to well above 2000 Hz.

The woofer's impedance rises from 4 Ω in the 200 to 300 Hz band to 6 Ω at 1200 Hz and 8Ω at 2000 Hz. For a 1200 Hz crossover, this would suggest a choke of 0.8 mH; but, allowing for a voice coil inductance of 0.3 mH, the right choke value turns out to be an inductance of about 0.5 mH.

The schematic and complete plans for this system are given in Chapter 15.

PROJECT 5: A COMPACT CLOSED BOX SPEAKER SYSTEM

The plans for the enclosure used in this project (Fig. 6-10) are based on a plan published by Phillips Gloeilampenfabrieken, the manufacturer of Norelco component loudspeakers distributed by Amperex. We have shown the correct English measurements to give a volume of 0.25 ft.^{3}, the equivalent of 7 liters.

Fig. 6-10. Project 5 completed.

The 1 in. tweeter has excellent dispersion, and the woofer is well damped. Initial measurements on the woofer in the box showed the final Q to be well under 0.7, which is the usual minimum, so that affected some of the decisions on the crossover network design. Phillips' original plans call for a 2.1 mH choke, an iron core inductor that they manufacture. That is the calculated value for this woofer, which has an impedance of about 19.5 Ω at 1500 Hz. Not having a choke with an iron core that would fit neatly into the small space, I decided to make up a choke, but an **air-core** choke of about 2 mH made with #18 gauge wire would be a bit bulky. Evidently Norelco had not subtracted the voice coil inductance from the indicated choke inductance, and one can assume that their choke worked out, well or they would have altered the value. A quick check suggested that if one allows for voice coil inductance, the choke value should be about 1 mH, which would be a much smaller choke. However, at all frequencies, a filter across the woofer will keep its impedance constantly around the minimum of 7.5 Ω that occurs at 200 Hz. The filter is a simple series RC combination, the 7.5 Ω resistor is just slightly greater than the voice coil's dc resistance of 7Ω, and the capacitor can be calculated from this formula:

$$C = \frac{0.001 \text{ H (approx. voice coil inductance)}}{(7)^2 \text{ (voice coil dc resistance squared)}}$$
$$= 0.00002 \text{ f } 20 \text{ }\mu\text{f}.$$

With this network wired across the woofer, its impedance at 1500 Hz is 7.5 Ω. Using that figure, the correct value for the choke turns out to be 0.8 mH.

Then, considering the low value the Q, I wound a choke from #22 gauge wire on the small coil form shown in the plans. This adds some resistance that one normally wants to avoid in a woofer circuit, but it raises the Q slightly, which is not a bad thing in this case. Both this coil and the one in the tweeter circuit can be made from a single roll of magnet wire. The values in the tweeter circuit are the same as those used by Phillips.

The system resonance of this speaker was 68 Hz, very close to the 70 Hz specification given by Phillips. Some reports indicate that the newest Norelco speakers have a stiffer suspension. If true, the system resonance with a new speaker would be higher, and the Q may also be higher. If your speaker has the model number 7066 instead of 7065, you should test the speaker and compare its performance to the one used in this project which was:

$f_s = 41$ Hz, $Q = 0.32$, $V_{as} = 0.9$ ft.^{3}

This system has exceptionally clean and wide range sound for its size. It benefits from a slight boost in the lower bass, and in most rooms a slight treble cut is appropriate, but the sound is alive and natural.

Project plans are in Chapter 15.

7 Reflex Systems

The reflex offers both a challenge and a promise. The challenge comes from the driver/enclosure interaction, which makes it far trickier than a simple closed box speaker; the payoff is greater bass range at lower distortion in the same size box.

The traditional design approach to ported enclosures has usually consisted of a little bit of planning and a lot of trial and error. When so much depended on luck, big boxes were the lesser evil, at least to the ear. The best advice then for any novice to follow was "the bigger the better." When a reflex sounded bad, most audio fans blamed the tuning, but flaws usually went deeper than a wrong port size. The bane of the boom box was that it had too many uncontrolled, and often unrecognized, variables such as driver Q. Even experts gave conflicting advice: Tune the box to the speaker's free air resonance, free air resonance means nothing, make the impedance peaks equal in amplitude, don't worry about the height of the peaks, make the ratios of the critical frequencies equal, and so on. Such confusion as this could make an amateur speaker builder feel like an orphan lost in a strange country where each and every inhabitant spoke a different language.

Now several people—notably Novak, Thiele, Small, Newman, Keele, and others—have laid a solid foundation under the shaky structure of bass reflex design theory. This has produced a revolution in reflex speaker box design that may affect the next generation of loudspeakers as much as the acoustic suspension revolution affected the last one.

One of the first people to challenge the "bigger is better" doctrine was James F. Novak, now Vice President of Engineering at Jensen. In 1959, he compared the performance of the compact closed box with that of the ported box and showed that a properly designed ported box, by damping the driver at resonance, produced less bass distortion. His analysis brought into focus the problem of using efficient speakers with large magnets in a closed box system, the same difficulty that was illustrated in our previous example of the 12 in. speaker with a Q of 0.31. The reflex means greater efficiency and reduced demand on the amplifier.

One of Novak's most significant contributions to ported box design was his optimum volume concept. He contended that for each driver, there is an optimum volume that will produce the smoothest and most extended bass response consistent with good transient response. Novak suggested a driver Q of 0.3 to 0.4, and he calculated that the optimum volume would occur when the ratio of driver compliance to box compliance (C_{ms}/C_{mb}) is equal to 1.44. Earlier recommendations had put this ratio at less than 1, usually about 0.5. Such low ratios produced the extremely large cabinets that dominated the hi-fi scene until the "acoustic suspension" revolution made them obsolete.

Novak's ratio did not always produce a small enclosure. He specified a low resonance speaker that naturally had greater compliance than earlier drivers, but he showed that if necessary, and if the tuning was correct, these high compliance woofers could be used in smaller than optimum volumes.

About the time that Novak proposed his concepts and developed a simplified method of computing the response of a ported system, A. N. Thiele, an Australian engineer, was trying to match a 12 in. woofer to a certain dictated cabinet volume. Thiele couldn't get the results he wanted, so he resorted to a closed box for the speaker, as many other engineers were doing. He saw Novak's paper and kept studying it until he realized that Novak's simplified analysis had exposed the ported box as being the equivalent of a high-pass electrical filter. Thiele applied network analysis to vented enclosures and came up with a slightly different value for Novak's optimum compliance ratio, 1.414 (instead of 1.44) for a speaker with a Q of precisely 0.383. He went further with his filter analogy and obtained a series of optimum conditions for drivers with Q's ranging from 0.18 to 0.6 and even greater. Thiele showed data for 28 different ways of making ported systems which he called "alignments."

Thiele classified his alignments according to the kind of electrical filter that would have the same kind of response curve near the low cut-off point. He labeled those with the flattest response Butterworth alignments; those with ripples in their response near cut-off, Chebyshev alignments. Using Thiele's alignment data, a builder can design a ported system that can vary in size from boxes with very low compliance (compared to that of the driver) and a bass cut-off higher in frequency than the driver's free air resonance, to large boxes with higher compliance than the driver, but with a cut-off frequency below the driver's free air resonance. He also developed a series of alignments in which there is either a roll-off, hump, or dip in the response curve which can be corrected by auxiliary electrical circuits. Thiele published his work in Australia in 1962, but it received little attention in the United States until it was republished here, about ten years later.

The first 9 alignments in Thiele's table require no special circuits. Figure 7-1 gives a summary of those alignments as rewritten by D.B. Keele, Jr. to include a new one which Keele designated "9.5" and added because of its low cut-off frequency. Thiele's original table had been presented from the point of view of the loudspeaker engineer who wants to design a driver to fit a certain kind of box. Keele has reworked the data to make it more useful for the person who has a certain driver and wants to design a box for it. Instead of stating a speaker compliance/box compliance ratio, Keele shows the ratio of the optimum box volume to the speaker's compliance in equivalent air volume: V_B/V_{as}. So by simply multiplying the value of the V_B/V_{as} ratio times V_{as}, you can get the correct box volume. Keele's chart also includes data on the ratio of the two impedance peak frequencies to the speaker's free air resonance, providing an additional check to see if the system is working right. In addition to the symbols used for closed box design, here are four new ones:

f_3 = the cut-off frequency: response down 3 dB
f_B = the tuned (resonance) frequency of the box
f_H = the frequency of the upper impedance peak
f_L = the frequency of the lower impedance peak.

HOW TO DESIGN A PORTED BOX FOR A DRIVER

Measure the values of f_s, Q, and V_{as} by Tests 1, 3, and 6 in Chapter 3. Next, find an alignment that fits the Q. It is assumed here that your amplifier will be a modern one that has a high damping factor. In that case, the Q_t for the speaker in the amplifier output circuit will be the same as the Q of the speaker alone.

Here is an example. Measurements on a specific speaker, a 10 in. woofer with a roll suspension and a 40 oz. magnet, show:

$$f_s = 32 \text{ Hz}$$
$$Q = 0.38$$
$$V_{as} = 5 \text{ ft.}^3$$

	ALIGNMENT DETAILS				BOX DESIGN				IMPEDANCE PEAK FREQUENCIES		
	NO.	TYPE	K	RIPPLE (db)	f_3/f_s	f_b/f_s	V_b/V_{as}	Q_t	f_1/f_s	f_h/f_s	f_h/f_1
QUASI-THIRD ORDER	1	QB3			2.68	2.000	.0954	.180	.5127	3.901	7.61
	2	QB3			2.28	1.730	.1337	.209	.5161	3.346	6.48
	3	QB3			1.77	1.420	.2242	.259	.5282	2.681	5.075
	4	QB3			1.45	1.230	.3390	.303	.5406	2.273	4.205
FOURTH ORDER	5	B4			1.000	1.000	.7072	.383	.5688	1.758	3.09
	6	C4	1.0		.867	.927	.9479	.415	.5771	1.607	2.78
	7	C4	.8		.729	.829	1.372	.466	.5741	1.445	2.52
	8	C4	.6	.13	.641	.757	1.790	.518	.5615	1.348	2.40
	9	C4		.25	.600	.716	2.062	.557	.5499	1.302	2.37
	9.5	C4		.55	.520	.638	2.60	.625	.5166	1.235	2.39

Fig. 7-1. Summary of QB3, B4, and C4 alignments from Thiele, rewritten by D.B. Keele, Jr. Reprinted with permission from the Journal of the Audio Engineering Society.

Figure 7-1 shows that this speaker is a natural for alignment #5, a B_4 (fourth order) Butterworth alignment. This is the classic Novak alignment and, when applicable, is one of the most useful. It permits flat response down to the free air resonance of the woofer.

For the 10 in woofer described above, the optimum volume box is $0.707 V_{as}$, or 3.5 ft.^{3} This would be the right volume if the system were perfect. Richard Small has found that the typical enclosure needs about 30% more volume if the predicted bass response is to be realized. That would make the practical volume for the 10 in. woofer about 4.55 ft.^{3} or, rounded off, 4.6 ft.^{3}

The tuning ratio is 1.000, so the box should be tuned to 32 Hz. The cut-off frequency/free air resonance ratio is also 1.000, which means that the low frequency response should cut off at 32 Hz. So the specifications from the chart for this speaker is a 4.6 cu. ft. enclosure that is tuned to 32 Hz.

The next question is, what kind of port? To find the area of the port and the length of duct behind it, we go to the design charts in Fig. 7-2 to 6. The area should be as large as possible, up to the effective piston area; it should not be so large that it would require a duct long enough to change the enclosure from a reflex into a tuned pipe. As a rule of thumb, the duct should be a few inches shorter than the depth of the box so that the rear opening will be unobstructed. You can use an L-shaped duct to get more length in a compact box, but for most enclosures of moderate size, a 5 in. to 8 in. duct has the advantage that you can use a simple straight tube.

To get some kind of figure for a minimum port area, we do Test 11 and find X_{max} to be about ⅛ in. and V_D to be 6.25 in.^{3} The minimum vent area then is:

$$S_V = (0.02)(32)(6.25)$$
$$= 4 \text{ in.}^2$$

For a safety margin, a port should have double the minimum area, or 8 in.^{2} in this case. The 3 in. tube has an area of 7 in.^{2}, which will be adequate if we need to use a small area port. There are no figures for a 4.6 ft.^{3} box in the charts, so we will look up figures for a 4 ft.^{3} box. By substituting a smaller volume in the chart, we will get a duct length that is slightly too long, but it is easier to cut off extra duct than to add more. The right length of 3 in. diameter tube for 30 Hz is 3 in. Since this is probably longer than necessary, we'll go to the next larger port.

Figure 7-4 shows data for a port that has twice the area of the 3 in. tube. For a 4 ft.^{3} box tuned to 30 Hz, it shows a duct length of 7⅜ in. This is a manageable length, so we will choose it. The 7⅜ in. duct will probably be longer than necessary because we have selected the length to tune a 4 ft.^{3} box to 30 Hz instead of a 4.6 ft.^{3} box to 32 Hz.

FREQUENCY (Hz)
LENGTH IN INCHES

VOL. (FT.^{3})	20	25	30	35	40	45	50	60	70	80	90	100
0.5						7	5⅜	3¼	2	1¼		
0.75					5⅝	4	3	1⅝	⅞			
1				5½	3⅞	2¾	1⅞	⅞				
1.25			6	4	2¾	1⅞	1¼					
1.5		7⅝	4¾	3⅛	2	1⅜	¾					
1.75		6¼	4	2½	1½	1						
2		5⅜	3¼	2	1¼							
2.5	7	4	2¼	1¼								
3	5⅝	3	1⅝	⅞								
3.5	4⅝	2½	1¼									
4.	3⅞	2	1									
5	2¾											

Fig. 7-2. Duct length for port with 3.14 in.^{2} of area (2 in. tube).

VOL. ($FT.^3$)	FREQUENCY (Hz) LENGTH IN INCHES 20	25	30	35	40	45	50	60	70	80	90	100
0.5								8⅜	5⅝	3¾	2½	1⅝
0.75						10¼	7⅞	4¾	3	1¾	⅞	
1					9⅝	7⅛	5⅜	3	1⅝	¾		
1.25					7¼	5¼	3⅞	2	⅞			
1.5				8⅛	5¾	4	2⅞	1¼				
1.75				6⅝	4½	3⅛	2⅛	¾				
2			8⅜	5½	3¾	2½	1⅝					
2.5			6¼	4	2½	1½	¾					
3		8	4¾	3	1¾	⅞						
3.5		6½	3¾	2¼	1¼							
4	9⅝	5⅜	3	1⅝	¾							
5	7¼	3⅞	2	⅞								
6	5¾	2⅞	1¼									
7	4½	2⅛	¾									
8	3¾	1⅝										
10	31/16	⅞										
12	2⅛											

USE SMALLER PORT

USE LARGER PORT

Fig. 7-3. Duct length for port with 7 in.^{2} of area (3 in. tube).

VOL. (FT.³)	20	25	30	35	40	45	50	60	70	80	90	100
0.5										8¾	6¼	4½
0.75										4¾	3⅛	1⅛
1								7⅜	7¼	2⅞	1½	
1.25								5⅜	4⅝	1⅝		
1.5						9⅜	9	3⅜	3	¾		
1.75					10⅜	7⅝	7	2⅞	2			
2				9¼	8¾	6¼	5½	2⅛	1⅜			
2.5			11	7¼	6⅜	4⅜	4½	2⅛	¾			
3		12	8⅛	5⅝	4¾	3⅛	3	1⅛				
3.5		9	7⅜	4⅝	3⅝	2½	2					
4		7	5¼	3	2⅞	1½	1¼					
5		5½	3⅞	2	1⅝							
6		4½	2⅞	1⅜	¾							
7	10¾	3	2⅛	¾								
8	8¾		1									
10	7⅜											
12	5½	1⅞										

FREQUENCY (Hz) / LENGTH IN INCHES

USE SMALLER PORT

USE LARGER PORT

Fig. 7-4. Duct length for port with 14 in.² of area (3 ¾ in. × 3 ¾ in.).

VOL. (FT.³)	20	25	30	35	40	45	50	60	70	80	90	100
0.5												9 3/8
0.75											7	2 5/8
1									9 5/8	10	4 1/4	1 1/4
1.25								8 3/8	6 7/8	6 3/8	2 1/2	
1.5							9 3/8	6 5/8	5	4 1/4	1 3/8	
1.75							6 5/8	5 1/4	3 3/4	2 7/8		
2					9 7/8	9 1/4	4 7/8	3 3/8	2 3/4	1 7/8		
2.5				11 5/8	7 7/8	7	3 5/8	2 7/8	1 3/8	1 1/8		
3			10 7/8	9 5/8	6 3/8	5 3/8	2 5/8	1 1/4				
3.5			8 3/8	6 7/8	4 3/8	4 1/4	1 1/4					
4			6 5/8	5	2 7/8	2 1/2						
5		11 1/4	5 1/4	3 3/4	1 7/8	1 3/8						
6		9 3/8	3 3/8	2 3/4	1 1/8							
7		6 3/4	2	1 3/8								
8		4 7/8										
10	11 1/4											
12	14 5/8											

FREQUENCY (Hz) / LENGTH IN INCHES

USE SMALLER PORT

USE LARGER PORT

Fig. 7-5. Duct length for port with 25 in.² of area (5 in. × 5 in.).

VOL. (FT.³)	20	25	30	35	40	45	50	60	70	80	90	100
0.5												11 7/8
0.75												7 3/8
1											10 1/2	4 3/4
1.25										10 3/4	7 1/4	3
1.5									9 5/8	8	5	1 3/4
1.75								8 7/8	7 3/4	6	3 1/2	3/4
2							9 3/8	6 1/2	5	4 1/2	2 3/8	
2.5						10 1/2	7 1/2	4 3/4	3 1/4	2 1/2	3/4	
3					10 3/4	7 1/4	4 3/4	3 3/8	1 7/8	1		
3.5				9 5/8	8	5	3	1 5/8	1			
4				7 5/8	6	3 1/2	1 3/4					
5				5	4 1/2	2 3/8	3/4					
6			8 7/8	3 1/4	2 1/2	3/4						
7			6 1/2		1 1/8							
8		11 7/8										
10												
12												

FREQUENCY (Hz)
LENGTH IN INCHES

USE SMALLER PORT

USE LARGER PORT

Fig. 7-6. Duct length for port with 49 in.² of area (7 in. × 7 in.).

For any combination that appears on the chart (that is, if we wanted to tune a 4 ft.³ box to 30 Hz), it is a good idea to add about 20% length; but if you have no test equipment, you should cut the duct to the exact length given by the chart.

When you calculate the dimensions to give the right internal volume, do not forget to add enough extra volume to account for the volume of the speaker, the duct, the cleats, and the bracing material in the box. As a rule of thumb, you can simply add 10% to the required volume and the net volume will usually be just about right. So for a net volume of 4.6 ft.³ we would built a box of about 5 ft.³ The exception to the 10% rule occurs with small boxes when a large area port has been chosen, which will require a longer than average duct.

In compact boxes, the damping material can change the effective volume from the calculated value. For one method of coping with volume shift, read the discussion of Project 8 in this chapter.

HOW TO TUNE A PORTED BOX

The easiest way to tune a ported box is to have the duct outside the box so that you will not need to remove the back every time you change the duct length (Fig. 7-7). If you are using a tubular duct, you can cut the hole for it large enough for the duct to be inserted in the speaker board. If the duct does not make an airtight seal with the

Fig. 7-7. The easiest way to tune a ported box is with the duct outside the box so you will not have to remove the back every time you change the duct length.

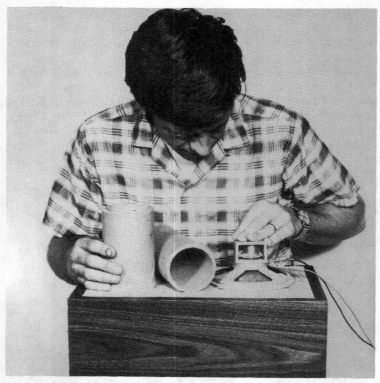

Fig. 7-8. Tuning a ported box by Test 7, Method IV: the listening method, using a cheap speaker.

board, you should tape the outside of the tube with a layer or two of masking tape to make a tight fit. If you are using a rectangular duct, line one end of the duct with foam weatherstripping tape and press the duct against the speaker board to make a seal between the duct and the board. For the rectangular duct, the cut-out in the baffle should be the same as the internal dimensions of the duct.

Although you do the tuning with the duct outside the box, there will be an almost insignificant shift of tuning when the duct is placed in the box unless the duct volume makes up a larger-than-average percentage of the enclosure volume. If there is a shift in tuning frequency, it will be slightly upward.

Sometimes it is convenient to tune an enclosure to a certain frequency before installing the speaker that will be used in it. For example, if you know the specifications of a speaker that you intend to buy or already have on order, you can go ahead and build the box and tune it using Method IV of Test 7 (Fig. 7-8).

To tune any enclosure, prepare a duct and measure f_B by Test 7. If you made the duct purposely too long, you will find that f_B is too low. To find out how much should be cut off its length, use this formula which is after Keele:

$$\Delta L_V = - \Delta f_B \frac{2 L_V}{f_B}$$

where:

ΔL_V = change in vent length in inches
Δf_B = required change in box frequency
L_V = length of duct when tested
f_B = frequency of box as tested

Notice that this formula gives a negative value when the vent is too long, showing how much to remove. If the vent were too short and you needed to add more length, the change in frequency (Δf_B) would be a negative number, and so the equation would give a positive value.

For the woofer above, suppose you put in a 7½ in. duct and find that f_B is 30 Hz. To find how much duct length to remove for a tuned frequency of 32 Hz:

$$\Delta L_V = - \ 2Hz \ \frac{15 \text{ in.}}{30 \text{ Hz}} = 1 \text{ in.}$$

The result indicates that you should cut 1 in. from the duct length, reducing it to 6½ in. long.

In this case you would make a duct that measures 3¾ × 3¾ in. inside, and 6½ in. from the front of the speaker board to the end of the duct. If there is no room on the board for a square port, you can make the port rectangular, but make it as nearly square, or as round, as possible. Narrow slots can be unpredictable; they add too much resistance to air flow. You can use plywood ducts in place of the tubular ones, but make sure they have the same cross sectional area as the circular tubes if you use the tuning charts. For square ducts to have about the same area, the length of one side of the square should be 1¾ in. for 2 in. tubes, or 2⅝ in. for a 3 in. tube.

In a properly designed and constructed reflex, the port output at resonance will exceed that of the driver by many times. Keep the port clear, or the damping on the driver will be impaired. Small vents are especially susceptible to loss of efficiency if they are partially blocked. Even grille cloth can affect the performance of a high velocity (small) vent and reduce speaker damping. You can test the effect of any material on the vent by stretching it over the vent while driving the speaker to high output at f_B. If the speaker shows an increase in movement when the material is placed over the vent, you should seek a more open material for the grille cloth. Keep damping material away from the interior of the vent unless you are designing a hybrid system. Too much damping material can also upset the calculated box volume by producing volumetric expansion, which makes the reflex system unpredictable. A 1 in. to 3 in. layer over the interior surfaces of the box near the woofer is usually enough.

HOW TO INTERPOLATE FROM THIELE'S ALIGNMENT DATA

As noted earlier, the B_4 alignment was a natural for the woofer with a Q of 0.38. It is a rare speaker that has a Q that is exactly right for one of Thiele's alignments. In the more usual situation, the value for Q will have to be interpolated between the values listed to find the tuning frequency and box volume. Here is the data on a 15 in. woofer:

$$f_s = 21 \text{ Hz}$$
$$Q = 0.34$$
$$V_{as} = 30 \text{ ft.}^3$$

Looking at the design chart, you can see that a Q of 0.34 lies between alignments #4, a third order quasi-Butterworth (QB_3) type, and #5, the familiar B_4 alignment. Note that the Q_T value for #4 is 0.303; for #5, 0.383. That makes a tabular difference of $0.383 - 0.303$, or 0.080. We know the Q of the woofer to only two places to the right of the decimal point, but to make interpolation easy, let's assume the precise value to be 0.343. At that figure, it would fall at exactly half the tabular difference between alignments #4 and #5. (Without laboratory-quality measuring equipment, this is as good as the testing accuracy anyway.) To find box volume, we have the following information:

For a Q_t of:

0.303, $V_B = 0.3390 \ V_{as}$
0.343, $V_B = ?$
0.383, $V_B = 0.7070 \ V_{as}$

Notice that a relatively small change in Q changes the ideal box volume over 100%. To get the ideal volume for the 15 in. woofer, we must find the tabular difference for the box ratio. That difference is $0.7070 - 0.3390$, or 0.3680. Since the woofer's Q lies at half this tabular difference for Q, the box ratio will also occur at half the tabular difference between the two ratios shown. Half the difference is 0.5×0.3680, or 0.1840. This figure, added to the value for alignment #4 (0.3390), gives a box ratio of 0.5230. So the optimum volume for this reflex system is $0.5230 \ V_{as}$. The 15 in. driver has a V_{as} of 30 ft.^{3}, so the box volume should be 0.5230×30, or about

15.7 ft.³ When we add 30% over volume, the final volume is 20.4 ft.³! This shows one of the disadvantages of using large woofers unless they are designed for use in boxes of moderate size.

To find the tuning frequency of this combination, you can perform the same kind of interpolation in the next column to the left. The tabular difference here is 0.230; the half point is 0.115. This shows up in the tuning ratio as 1.115. The tuning frequency (f_B) then should be 1.115 × 21, or 23.4 Hz. The cut-off frequency can be found by the same method, and it turns out to be about 25.7 Hz, or say, 26 Hz.

USING A POCKET CALCULATOR TO DESIGN LOUDSPEAKER ENCLOSURES

A pocket calculator does make quick work of the arithmetic in speaker system design work, but it can do more than that. I am indebted to D. B. Keele, Jr., Chief Engineer at Klipsch and Associates, and formerly with Electro-Voice, who has worked out a method of using a pocket calculator to design a ported box without having to refer to charts.

Keele's Pocket Calculator Method of Vented Box Design

It takes two flow charts to explain the idea behind this method of vented box design. Fig. 7-9 shows the general procedure; Fig. 7-10 illustrates the catchy part. If no changes need to be made in the box size, you use the first set of equations for f_3 and f_B; if any changes are made, you use the second set.

General Form

Given: Driver Thiele/Small Parameters

1. f_s = the frequency of free air resonance
2. Q = the total driver Q, *sometimes* labeled "Q_{ts}"
3. V_{as} = the compliance equivalent volume

Find:

1. V_B (ft.³), the net box volume
2. f_B (Hz), box frequency of resonance
3. f_3 (Hz), the system cut-off frequency, down 3 dB
4. Hump (or dip) in passband

Procedure:
$$V_B = 15\ Q^{2.87} V_{as}$$

If no change in box size:

1. $f_3 = 0.26\ Q^{-1.4}\ f_s$
2. $f_B = 0.42\ Q^{-0.9}\ f_s$
3. The design is optimum. There is no hump or dip.

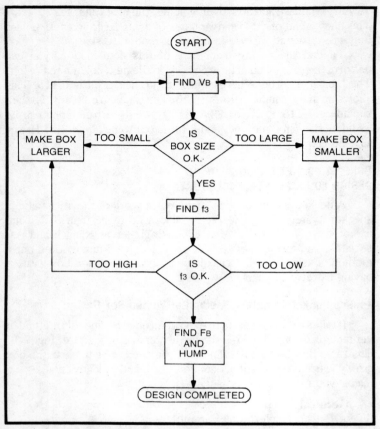

Fig. 7-9. Oversimplified flowchart of Keele's Pocket Calculator Method of Vented Box Design.

If any change is made in box size:

1. $f_3 = \left(\sqrt{\dfrac{V_{as}}{V_B}}\right)(f_s)$

2. $f_B = \left(\dfrac{V_{as}}{V_B}\right)^{0.32}(f_s)$

3. Hump (dB) $= 20 \log \left[2.6\, Q \left(\dfrac{V_{as}}{V_B}\right)^{0.35}\right]$

Worked Example:

Given:

15 in. woofer (same driver used for second example of chart design)

$$f_s = 21 \text{ Hz}$$
$$Q = 0.34$$
$$V_{as} = 30 \text{ ft.}^3$$

Procedure:

$$V_B = 15 \, (0.34)^{2.87}(30)$$
$$= 20.4 \text{ ft.}^3$$

If no change in the box size:

$$f_3 = 0.26(0.34)^{-1.4}(21)$$
$$= 24.7 \text{ Hz}$$

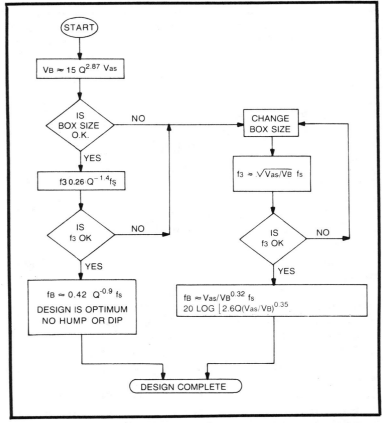

Fig. 7-10. Simplified flowchart of Keele's Pocket Calculator Method of Vented Box Design.

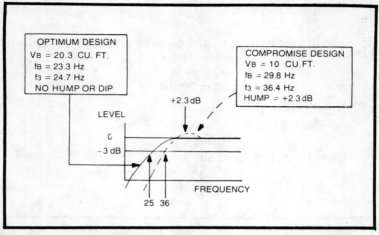

Fig. 7-11. Graph of values obtained for worked example of Keele's Pocket Calculator Method of Vented Box Design.

$$f_B = 0.42 \,(0.34)^{0.9}(21)$$
$$= 23.3 \text{ Hz}$$

If box size is too large try:
$$V_B = 10 \text{ ft.}^3$$
$$f_3 = \sqrt{30/10}\,(21)$$
$$= 36.4 \text{ Hz}$$

If f_3 is satisfactory:
$$f_B = (30/10)^{0.32}(21)$$
$$= 29.8 \text{ Hz}$$
$$\text{Hump} = 20 \text{ Log }[2.6(0.34)\,(30/10)^{0.35}]$$
$$= +2.3 \text{ dB}$$

Figure 7-11 shows a graph comparing the frequencies of these two different enclosures.

One advantage of the hand calculator method is that the math goes so quickly, you can investigate several values for cubic volume and easily determine the kind of response, tuning, and bass cut-off frequency for each. This exercise will make you aware of the kind of compromise decisions faced by loudspeaker engineers who must consider such practical questions as how large an enclosure the customer will tolerate. The values obtained by following the alignment chart are the best guide for the flattest possible frequency response. You can reduce cabinet volume if you also are willing to accept a more restricted bass range, and if you take the precaution to change the tuning ratio to the one derived from Keele's program.

The ideal type of calculator to use for this purpose is a "scientific" model, such as a Hewlett Packard 21, 25, 35, 45, 65, etc., or a Texas Instruments SR 50, 51, 52, 56, 60, etc., or any other brand of calculator with similar functions. You can get by with a much simpler model, but it should have at least a ($\sqrt{x}$) function, and you may need to use log tables with it. The ($\sqrt{x}$) function can be used to approximate the y^x problem. To check the accuracy of one of these calculators on a vented box problem, I compared the optimum V_B obtained both by the simple calculator method and by the much longer paper-and-pencil-plus-logarithms way, which is theoretically more accurate. The answers were 20.307915 ft.^{3} for the calculator, and 20.353725 ft.^{3} for the drudgery method. Not much to complain about, considering that the calculator took only a small fraction of the time spent with paper and pencil. Of course, the more sophisticated calculators will do every part of the calculations, and do them more accurately and quickly. This problem was shown just to prove that you can use a less expensive calculator if you have one with a ($\sqrt{x}$) function. Read the instruction book that came with your calculator to see how to proceed.

Keele's Calculator Equation For Vented Box Frequency Response

This is another useful mathematical model that goes quickly with a hand calculator.

General Form

Given:
1. driver parameters
   a. f_s (Hz) = free air resonance
   b. Q = total driver Q, sometimes labeled "Q_{ts}"
   c. V_{as} (ft.^{3}) = compliance equivalent volume
2. box parameters
   a. V_B, net box volume
   b. f_B, box frequency of resonance

Compute:
1. Constants
   a. $A = f_B^2/f_s^2$
   b. $B = A/Q + f_B/(7f_s)$
   c. $C = 1 + A + f_B/(7f_sQ) + V_{as}/V_B$
   d. $D = 1/Q + f_B/(7f_s)$
2. Response in dB (Normalized to 0 dB at high frequencies)
   a. At each frequency f(Hz) Find: $f_n = f/f_s$
   b. Substitute into:

Response dB = 20 Log $\dfrac{f_n^4}{\sqrt{(f_n^4 - Cf_n^2 + A)^2 + (Bf_n - Df_n^3)^2}}$

Worked Example:
Driver is Electro-Voice EVM 18B

Given:
1. driver parameters
   a. $f_s = 33$ Hz
   b. $Q = 0.35$
   c. $V_{as} = 17.9$ ft.^{3}
2. box parameters
   a. $V_B = 6.5$ ft.^{3}
   b. $f_B = 45$ Hz

Compute:

Find response level at 45 Hz (down from mid-band)

1. Constants
   a. $A = (45/33)^2$
      $= 1.86$
   b. $B = (1.86/0.35) + 45/7(33)$
      $= 5.31 + 0.19$
      $= 5.51$
   c. $C = 1 + 1.86 + 45/7(33)0.35 + 17.9/6.5$
      $= 6.17$
   d. $D = 1/0.35 + 45/7(33)$
      $= 3.05$

Response in dB at 45 Hz:

$$f_n = f/33$$

For $f = 45$ Hz:

$$f_n = 45/33$$
$$= 1.36$$
$$f_n^2 = (45/33)^2$$
$$= 1.86$$
$$f_n^3 = (45/33)^3$$
$$= 2.54$$
$$f_n^4 = (45/33)^4$$
$$= 3.46$$

Response dB = 20 Log

$$\frac{3.46}{\sqrt{[3.46 - 6.17(1.86) + 1.86]^2 + [5.51(1.36) - 3.05(2.54)]^2}}$$

$= 20$ Log

$$\frac{3.46}{\sqrt{37.9 + 0.064}}$$

$= 20$ Log 0.56

$= -5.0$ dB (computor program printout shows: -4.9 dB)

Keele's Pocket Calculator Method of Computing Vent Dimensions

You can also use a calculator to find the right vent area/length ratio to tune your box to an exact frequency without using a chart. Once you find the ratio, you must juggle area vs length to obtain a vent that is neither too small in area nor too long for your speaker and box dimensions.

General Form

Given:

1. V_D = driver peak displacement volume
2. V_B = net box volume
3. f_B = box frequency of resonance

Compute:

1. Minimum vent area, $S_{V(min)}$ (based on large signal power output considerations to minimize vent noise)
   $S_{V(min)} = 0.02\ f_B\ V_D$ (after R. Small)
2. Vent area-length combinations (S_v, L_v)
   a. compute α (vent area/vent effective length)
   $\alpha = V_B\ (2\pi f_B/c)^2 \approx 3.7 \times 10^{-4} V_B f_B^2$
   where V_B is in ft.^{3} and f_B is in Hz
   b. compute L_v
   $L_v = S_v/\alpha - 0.83\sqrt{S_v}$

Start at $S_{V(min)}$ and find L_v, step up with 2 $S_{V(min)}$, 4$S_{V(min)}$ etc. Select S_v and L_v so that L_v will not be too close to the back panel of the box. Good values would be the area and length combination corresponding to:

$$S_v = 2\ S_{V(min)}$$

if this is possible.

Worked Example:

Driver is Electro-Voice model EVM 18B

Given:

1. $V_D = 31$ in.^{3}
2. $V_B = 6.5$ ft.^{3}
3. $f_B = 45$ Hz

Compute:

1. $S_{V(min)} = 0.02\ (45)\ 31$
   $= 27.9$ in.^{2}
2. a. $\alpha = 3.7 \times 10^{-4}\ (6.5)\ (45)^2$
      $= 4.9$ in.^{2}/in.
   b. $L_v = 27.9/4.9 - 0.83\ \sqrt{27.9}$
      $= 1.3$ in.

NOTE: Lv may come out negative; step up Sv until $L_v \geq 0.75$ in. (for ¾ in. material).

Vent Length (Lv)	Vent area (Sv)
1.3 in.	27.9 in.^{2} $S_{V(min)}$
5.2 in.	55.8 in.^{2} 2 $S_{V(min)}$
13.9 in.	111.5 in.^{2} 4 $S_{V(min)}$

The best choice is the 5.2 in. vent length with 55.8 in.^{2} vent area.

D. B. Keele, Jr., who worked out the vented box design equations for the pocket calculator, mentions that this program will work well with speakers with Q's up to about 0.6. When it is used with speakers of higher Q, the box volume will be huge and the tuning low; the resulting systems will have extended flat response, but high distortion except at low sound levels.

It appears that some of the arguments about reflex tuning arose because different people were testing different kinds of speakers and trying to reach an impossible agreement on which was the best way to tune an enclosure to their speakers. For example, some experimenters have found that woofers with very low resonance, say below 20 Hz, work better in a box tuned to about 35 Hz. Normally, such low resonance speakers have high compliance cones and a low Q. The alignment chart shows that this kind of woofer can work well in a small enclosure that is tuned to a frequency well above the speaker's free air resonance. So theory has at last proved the experimenters right.

EQUALIZED REFLEX SYSTEMS

The design methods outlined earlier were set up to produce a woofer-enclosure system with flat response down to the bass cut-off point. If a reflex is tuned to a lower frequency, the response will droop above the box frequency as shown by curve B in Fig. 7-12, but if we add an equalizer to the amplifier circuit that peaks the signal just above the box frequency as shown in curve C, the complete system will have a flat, extended bass range, curve D. The equalizer costs something, but if you have a powerful amplifier and like to drive your speakers to rock concert levels, it can save your woofers. In addition to boosting the usable bass, the equalizer cuts off subsonic signals such as turntable rumble that can overdrive the speakers. Notice that a properly designed equalizer will deliver bass boost just above box resonance where the port damps the cone, so that the system can produce robust bass without distortion.

Ideally such an equalizer should be designed into the amplifier circuit by the manufacturer. The circuit should be adjustable so that

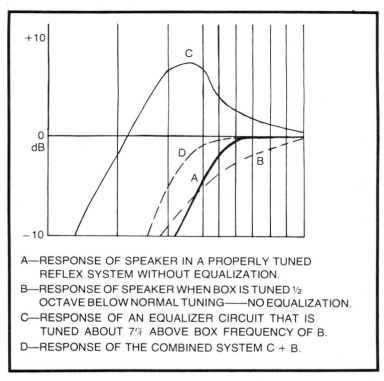

A—RESPONSE OF SPEAKER IN A PROPERLY TUNED REFLEX SYSTEM WITHOUT EQUALIZATION.
B—RESPONSE OF SPEAKER WHEN BOX IS TUNED ½ OCTAVE BELOW NORMAL TUNING——NO EQUALIZATION.
C—RESPONSE OF AN EQUALIZER CIRCUIT THAT IS TUNED ABOUT 7% ABOVE BOX FREQUENCY OF B.
D—RESPONSE OF THE COMBINED SYSTEM C + B.

Fig. 7-12. How an equalizer extends the response of a ported box speaker.

the frequency of the boost could be tuned to individual speaker systems. Some amplifiers do contain high pass filters that attenuate subsonic signals, and such amplifiers are a good choice for driving unequalized reflex systems. These unequalized and unprotected speakers should never be used with a combination of high powered amplifiers and poor turntables. Of course, this kind of a marriage makes no sense, but according to Murphy's Law, that is exactly why it happens.

Since amplifier manufacturers do not provide tuned equalizers, you will have to add one to your system if you want to get the extra bass response and low distortion they provide. One of these outboard equalizers can be hooked into the Tape Monitor circuit of your preamp. The design procedure would be:

1. Design a ported box, either by one of the first nine of Thiele's alignments, or by Keele's hand calculator method.
2. Tune the box 1/2 octave below the frequency indicated by Step 1.

3. Design an equalizer with a second order high-pass filter which peaks at 7% above the new box resonance.
4. Experiment with the box and equalizer design, if necessary, to get optimum performance.

To simplify Steps 1 and 2, you can use these formulas from D. B. Keele:

$$V_B = (4.1)(Q)^2 (V_{as})$$
$$f_B = 0.3 f_s/Q$$

Many unequalized systems cut-off down to 50 Hz or lower, so an equalized system should cut off at a lower frequency, say 20 or 25 Hz up to 50 Hz. Any speaker that has an F_s/Q ratio in the 80 to 160 range is a good choice for this kind of system.

If you want to try your hand at equalized reflexes, but you do not want to have to design an equalizer, you can build up one designed by Pat Snyder of Speakerlab. The schematic diagram is shown in Fig. 16-3. You can tune the circuit to any suitable frequency by controlling the values of C_2 and C_3. These values should be selected to give a cut-off frequency, f_{aux}, that is equal to the lowered box frequency when:

$$C_1 = C_2 = \frac{23.2}{f_{aux}}$$

Theoretically, the equalizer should provide a boost of 6 dB just above the cut-off frequency, but Snyder's equalizer has a variable resistance, R_4, that permits you to experiment with different amounts of boost up to 10 dB. This is useful if you use the equalizer with a speaker system that was not designed specifically for an equalized circuit. But if you design a reflex enclosure by the rules above, you can get a 6 dB boost by substituting a 30Ω resistor for R_4, a 970Ω resistor for R_6, and connect the lead from R_2 to the junction of R_4 and R_6.

Snyder has these suggestions for using the equalizer with any typical reflex system: Take a bass reflex system that already performs well and hook it up to the filter. Or even just put a likely looking speaker in a fairly plausible box and adjust the vent until it sounds best. Then hook up the bass reflex equalizer. For a start you might use $C_2 = C_3 = 0.5 \mu f$ so $f_{aux} = 46$ Hz. Run the peak height up and down and see what happens. If this has no effect on the sound, f_{aux} is too low—raise it. If varying the peak height definitely affects the sound but does nothing to improve it, f_{aux} is probably too high—reduce it. If large peak heights make the bass overstrong, but solid, you're in the ballpark. Reduce the peak height to where you get maximum realistic bass. When you get it set right, bass will go low and be solid, and the visible cone motion will be substantially reduced over what it is without the filter.

Snyder says that the equalizer completely eliminated cone surging produced by an FM tuner on station capture by cutting off subsonic frequencies—and yet produced a more powerful bass in the program material. He says that it adds ambience, a vast hall effect, either by reproducing low frequency reverberation, or perhaps by reducing the low bass distortion.

DESIGNING AN EQUALIZED REFLEX SYSTEM: A WORKED EXAMPLE

After choosing a suitable speaker, make the appropriate tests to determine f_s, Q, and V_{as}. For this example, we will choose an 8 in. Speakerlab woofer, W804R, with a f_s/Q ratio of about 84. This value indicates that the cut-off frequency, which will be at about 0.3 f_s/Q, will fall around 25 Hz. We would not want to set the cut-off much lower because subsonic energy could then damage the woofer.

The values for the W804R are:

$$f_s = 26.1 \text{ Hz}$$
$$Q = 0.31$$
$$V_{as} = 3.7 \text{ ft.}^3$$

To find the optimum volume:

$$V_B = (4.1)(0.31)^2(3.7)$$
$$= 1.46 \text{ ft.}^3$$

Adding 10% to account for space occupied by the speaker, the duct, and bracing:

$$V_B \text{ (total)} = 1.46 \text{ ft.}^3 + 0.146 \text{ ft.}^3$$
$$= 1.6 \text{ ft.}^3$$

And the tuned frequency of the box should be:

$$f_B = \frac{(0.3)(26.1)}{0.31} = 25 \text{ Hz}.$$

So the box should have a total of 1.6 ft.³ and be tuned to about 25 Hz, requiring a 7 in. length of tube with a diameter of only 2 in. This combination would require either enough cabinet depth so the duct would not extend too close to the rear wall, or an L-shaped duct; but more important, at this low frequency, a port of such small area might produce an air velocity so high that any grille cloth would reduce vent output. The best solution would be to use a tube with a smooth interior, such as a length of plastic pipe, and place it where it would not be covered by the grille. Another possibility would be to use a passive radiator and add enough mass to the cone to tune the system to 25 Hz. For high power applications, a better solution would be to use a speaker that would require a larger enclosure, such as the Speakerlab W1504S. Its specifications are:

$$f_s = 18.9 \text{ Hz}$$
$$Q = 0.201$$
$$V_{as} = 31.6 \text{ ft.}^3$$

For this speaker:

$$V_B = (4.1)(0.201)^3(31.6)$$
$$= 5.23 \text{ ft.}^3 + 10\% = 5.8 \text{ ft.}^3$$

And:

$$f_B = \frac{(0.3)(18.9)}{0.201} = 28 \text{ Hz}$$

This 5.8 ft.³ box can be tuned to 28 Hz by a 3¾ × 3¾ in. duct that is about 5 in. or 6 in. long, which is a practical length and, more important, has about 4½ times the area of the tube in the smaller box. Of course, the peak displacement volume of the 15 in. speaker may be considerably greater than that of the 8 in. model, so again it may be desirable to avoid any covering over the vent. If a grille cloth is found to be too restrictive, the vent can be located in the back or bottom of the box. If the vent is placed in the bottom, the legs or feet under the enclosure can be designed to raise the box just high enough to tune it. This can be done by locating the port in the center of the bottom and experimenting with various lengths of feet until the box is tuned to the right frequency.

To use the Speakerlab equalizer with one of these systems, you can choose the values of C_2 and C_3 to satisfy the formulas mentioned earlier.ABy 23.2 by the cut-off frequencies, which would be 25 Hz for the 8 in. system and 28 Hz for the 15 in. model, the values of C_2 and C_3 would be about 0.93 and about 0.83. These values could be approximated by connecting standard value capacitors in parallel.

If you want to experiment with one of these equalizer-assisted speakers, check the plans and other information in Chapter 16.

HOW TO DESIGN A PASSIVE RADIATOR REFLEX

Although a reflex system can offer lower distortion and greater bass range than a closed box system of the same size, it is hard to tune a small box to a low frequency without creating high air velocity through the small port. If the vent is enlarged enough to avoid these effects, the duct length may be too great for the small box.

One solution is the passive radiator (Fig. 7-13). H. F. Olson described a "drone cone" in 1955, but it drew little attention until the late 1960's. The passive radiator can be tuned to a low frequency by adding mass to the diaphragm. Except for the different ways in which mass is added to the vent air (ordinary reflex) or diaphragm (passive radiator), the two types of reflex systems operate under the same principles.

Fig. 7-13. The passive radiator on the left, is a speaker without a magnet or voice coil.

For successsul reflex operation, a passive radiator (PR) should have high compliance. If it has the same diameter as the driver, the passive radiator should have twice the mass of the driver cone and about twice the compliance. The PR unit should be capable of displacing twice as much air volume as the driver. The air volume displacement of a vibrating diaphragm is the volume of air moved by the diaphragm at its maximum throw, equal to the area of the cone times the distance the cone moves.

In order to fulfill the volume displacement requirement, it is simpler to use a diaphragm that is larger than the driver. A 10 in. PR would be a better choice for an 8 in. woofer, or a 12 in. PR for a 10 in. woofer, than would be a PR cone the same size as the woofer. If the volume displacement ability of the PR unit is too low, it will not damp the woofer enough at resonance.

To design a PR system, use the same procedure as for an ordinary ported box. The only difference is that you must add some kind of mass to the PR unit, preferably at the rear of the cone's center part, until the tuned box frequency is low enough. One way to temporarily add mass for tuning is to use blobs of modeling clay for the added mass. Then, when the frequency is right, use a ruler balance to weigh an equal mass of other material to glue to the cone. Do not glue it all to the cone at once; glue about half the added mass on the cone and run another test. Then add more in small amounts until the box is tuned. The reason for adding the mass in small doses is that the glue itself also has mass, making the tuning too low in frequency if you are not careful.

To check the tuning of a PR system, feed a 5-10 V signal to the woofer, as in a ported system, and watch both the drone cone and the driver. The action of the drone should peak at the box frequency, and driver vibration should reach a minimum at that point.

Fig. 7-14. Project 6 in use.

One advantage of the PR system is that you can subtract a bit from the box volume and still get good results. Some PR systems operate well at a third or more lower volume than the theoretical ideal.

While PR systems have the advantage of working well in a small box, they have no advantage over ordinary reflexes unless there is a tuning problem in the ported box. Some PR proponents claim that they block mid-range reflections from inside the box, but those reflections can usually be damped with fiberglass or other damping material without over-damping the system or blocking the port.

PROJECT 6: ELECTRO-VOICE MC8A PORTED SYSTEM

Electro-Voice now publishes the specifications required to calculate speaker/enclosure performance by the Thiele-Small alignment data, so we will use an EV speaker for this project (Fig. 7-14). The MC8A is EV's lowest-priced speaker, but it has a die-cast frame and a frequency response ± 6 dB of 42 to 20,000 Hz in a 2.4 ft.^{3} vented enclosure. The response in that box is down 3 dB at 50 Hz.

The MC8A's critical specifications for box design are:

$$f_s = 75 \text{ Hz}$$
$$V_{as} = 1.05 \text{ ft.}^3$$
$$Q = 0.6$$

Using Keele's pocket calculator method to design a box for the MC8A, we find:
$$V_B = 15\ (0.6)^{2.87}\ 1.05$$
$$= 3.6\ ft.^3$$

Its cut-off frequency, where the response would be down 3 dB, is:
$$f_3 = 0.26\ (0.6)^{-1.4}\ 75$$
$$= 40\ Hz$$

And for this performance the box should be tuned to:
$$f_B = 0.42 \times (0.6)^{-0.9}\ 75$$
$$= 50\ Hz$$

So, according to these calculations, a 3.6 ft.³ box tuned to 50 Hz would give the most extended bass with an MC8A speaker, the response being down 3 dB at 40 Hz. This particular alignment permits bass extension well below the speaker's free air resonance, and even below the tuned frequency of the box. However, to demonstrate how to arrive at an alternate design, let's say we want a slightly smaller box. If we aim for a 3 ft.³ box, how much bass range will be lost?

The flow frequency cut-off will be:
$$f_3 = \sqrt{\frac{1.05}{3}} \times 75$$
$$= 44\ Hz$$

The box tuning frequency should be:
$$f_B = \frac{(1.05)^{0.32}}{3} \times 75$$
$$= 54\ Hz$$

Since we cannot expect to get a flat response down to cut-off by reducing the size of the box, we will check to see how great a hump this will make in the response curve.

$$\text{Hump} = 20 \log \left[2.6\ (0.6) \left(\frac{1.05}{3} \right)^{0.35} \right]$$
$$= .67\ dB$$

A peak of this magnitude is insignificant.

These calculations show that although there may be an ideal box design for any speaker, many other designs are also possible, depending on your choice. Which do you want more: a flat bass response or a compact cabinet? One big advantage of using the alignment data is that we can investigate the range and response characteristics of a certain speaker/box combination without having to build it and test it. Anyone who wants to build a box for the MC8A can choose between a box of moderate size or optimum volume, or any intermediate size such as the 2.4 ft.^{3} box suggested by Electro-Voice.

For this project we will choose the 3 ft.^{3} box and go through the steps of planning the enclosure to meet the required specifications.

Electro-Voice lists the peak displacement volume of the MC8A as 1.9 in.^{3}. So the minimum port area is:

$$S_V = \sqrt{0.02\ (61)\ (1.9)}$$
$$= 2.3\ in.^2$$

If possible, we will use an area that is at least double the minimum, so any port with an area of 5 in.^{2} or more will do. Going to the charts, we find that area is no problem for a 3 ft.^{3} box tuned to 54 Hz. Figure 7-5 shows that a 5 in. × 5 in. port (10 times the minimum area) that is 4⅞ in. long will tune a 3 ft.^{3} box to 50 Hz.

However, since Electro-Voice specifies that calculated enclosure volume should not include the volume displaced by speaker, port, and bracing, we will try to estimate the actual volume of the duct and speaker. If the 5 in. × 5 in. × (about) 4⅞ in. port is made of ½ in. plywood, its external cross sectional area will be 6 in. × 6 in. The front board thickness makes up ¾ in. of the 4⅞ in. length, so the duct will extend about 4 in. into the cabinet. The duct volume is 6 in. × 6 in. × 4 in. = 144 in.^{3}, or about 0.08 ft.^{3} The estimated volume correction should be about 0.23 ft.^{3}, 0.15 ft.^{3} for the speaker and 0.08 ft.^{3} for the duct. The bracing and cleats will occupy very little space, so a box of 3.25 ft.^{3} (5600 in.^{3}) should do. The cube root of 5600 is 17.75, so the width will be 17¾ in. Using the Fibonacci series, the depth will be 11 in. (17.75 × 0.62), and the height will be 28¾ in. (17.75 × 1.62). These are internal dimensions, so if we use ¾ in. material, the outside measurements will be 19¼ in. × 30¼ in. × 12½ in.

This concludes the planning stage of this project; the construction details are in Chapter 15.

PROJECT 7: JBL PORTED SYSTEM

Here we will use an expensive full range speaker in an enclosure that will permit the speaker to give its best performance (Fig. 7-15). The speaker is a JBL LE8T. You are not likely to see this speaker in an audio showroom unless it is in a compact box. Although the speaker performs well in compact enclosures, you can build a box that will give a better and more extended bass response.

JBL sells a box of specifications for enclosures, called an "enclosure kit," that will give you specific recommendations for housing JBL speakers, plus a useful book on enclosure design and construc-

Fig. 7-15. Project 7.

tion. After getting one of these kits, I decided I would design my own box for the LE8T.

Some quick tests on an LE8T, which is several years old but identical to current models, showed these figures:

$$f_s = 44 \text{ Hz}$$
$$V_{as} = 1.7 \text{ ft.}^3$$
$$Q = 0.5$$

Putting these values through my pocket calculator, I came out with these recommendations:

$$V_B = 3.5 \text{ ft.}^3$$
$$f_3 = 30 \text{ Hz}$$
$$f_B = 35 \text{ Hz}$$

Then I went to JBL's recommendations. They give specifications for enclosures in five ranges of volume in cubic feet: 0.75 to 1.0; 1.1 to 1.5; 1.6 to 2.0; 2.1 to 3.0; and 3.1 to 4.0. From the specifications for the ducts on the largest enclosure, it appears that they are tuned to the 35 to 40 Hz range from a 3.5 ft.^{3} box.

I suspect that JBL's recommendations are made on the basis of extensive testing, so it is interesting that the top volume that they list for this speaker is approximately that of the theoretical ideal volume. After all, there is no use in building a big cabinet just to have a big cabinet, particularly when the bigger box will give poorer performance.

Looking through their material, I found detailed plans for many enclosures, including a 3.0 ft.^{3} box for the LE8T. At first I ignored it; after all, why save a half cubic foot and risk losing some bass response? But then I applied the figures from the LE8T to my calculator to see how much bass range would be sacrificed:

$$f_3 = \sqrt{\frac{1.7}{3}} \quad (44)$$
$$= 33 \text{ Hz}$$

Of course, this reduction in box size may produce a hump in the low frequency response curve above the cut-off frequency. Checked out by formula and the measurements made on an LE8T, this hump turns out to be about 0.5 dB, which is too insignificant to be heard. Assuming that the measurements were reasonably accurate, these figures—a 3 Hz loss of range and a 0.5 dB variation in response—are remarkably close to the theoretical ideal, so close, in fact, that there seems little point in going to a larger box. When the grille cloth is removed, some people say the speaker looks too small for even this size box. This is one reason a homemade enclosure can outperform a factory job: too many customers will not accept the idea of a full size enclosure for a small speaker. They buy a compact and expect to get the same performance from it that a larger box would give; but many enclosures are a compromise between the best sound and other considerations, just as some of those in this book are compromises, but here we have mentioned the compromises and the reasons for them so that you can make your own decision about which way to go. The calculations made for this project, although rough approximations, show that when you rely on a manufactuerer's recommendations and he lists several sets of possible enclosure volumes for a speaker, you should choose the largest size he recommends, if possible. No manufacturer is likely to recommend an enclosure too large for his speaker.

The most interesting and common reaction to the two 8 in. ported box systems described in Projects 6 and 7 was summed up by one critical listener, an electrical engineer, who said, "They don't *sound* like 8 in. speakers."

PROJECT 8: A COMPACT PORTED BOX SPEAKER

The preceding projects illustrate that ported enclosures must usually be rather large relative to the speaker size in order to give

Fig. 7-16. Project 8 (front view without grill).

optimum performance. In this project, we will use a much smaller speaker, but one having a low Q, to get satisfactory performance in a small box (Fig. 7-16). The speaker is an Olson SP-245, a 5 in. × 7 in. model with a foam roll surround and a 16 oz. magnet. Notice that this is the actual magnet weight: the magnet *assembly* weighs 2½ lb., which is unusual for a 5 in. × 7 in. speaker. The speaker is sold as a car speaker, but car speakers often make good extension speakers or serve well for other purposes, such as upgrading a portable stereo system.

Speaker tests gave the following data:

$$f_s = 57 \text{ Hz}$$
$$V_{as} = 0.95 \text{ ft.}^3$$
$$Q = 0.3$$

The low Q looks good for a compact box design. A lower Q does not necessarily mean a better speaker, but if you want to put a speaker into a compact ported box and get a flat response without auxiliary circuits, a low Q is necessary. The cut-off point will be higher than it would have been for the same speaker with a higher Q in a larger box. If you put a low Q speaker in a box that is too large, the speaker will be overdamped. In this case, we will aim for a compact box to fit the speaker so that the response will be as flat as possible.

Applying Keele's calculator method of enclosure design, we find that the ideal volume for the speaker is 0.45 ft.^{3} or 788 in.^{2} Using the "golden" ratios and adding some volume, it appears that a box with internal dimensions of 9½ in. × 15½ in. × 6 in. will have a cubic volume of about 883 in.^{3}, an overvolume of about 13½% that should be close enough to the ideal volume.

If we have calculated correctly and closely guessed the amount of overvolume to provide, this box and speaker should be tuned to 70.7 Hz and have a low frequency cut-off at 80 Hz. This seems reasonable for a box much smaller than the speaker's V_{as}; in such cases, the cut-off point is always above the speaker's free air resonance. However, for a compact box, a small error in estimating speaker volume or port volume can upset the tuning much more than for a large enclosure. How can we know exactly how much volume to allow? Since we cannot be sure of the precise figure, we will use a slightly different approach to ported box design that will automatically tell us how to tune the box. We will build the cabinet and install the speaker, the port, and the damping material; *then* we will measure the V_{as}/V_B ratio, which can be used to precisely determine the best tuning frequency for this box. Before we measure the volume ratio, first we must tune the box to approximately the right frequency.

For such a small box, we will have to use a small area duct to tune it, so we go to the tuning chart for a 2 in. diameter tube and see that a 2 in. length of tube will tune a box of 0.5 ft.^{3} to 70 Hz. This box is smaller, so we will start with the full length of a frozen orange juice can whose ends are removed. Although the paper tube is thin-walled, it can be reinforced by tape; a curved surface is more rigid than flat material of the same thickness. We will use the full 3½ in. length of the tube because the box does not have to be tuned precisely in order to get the volume ratio, and the tube can be shortened later for more accurate tuning.

With the speaker installed and the walls covered with fiberglass, the critical frequencies are: f_L, 29 Hz; f_B, 58.5 Hz; and f_H, 102.5 Hz. Using these values, we can get the V_{as}/V_B ratio from:

$$V_{as}/V_B = \frac{(f_H + f_B)(f_H - f_B)(f_B + f_L)(f_B - f_L)}{(f_H)^2 (f_L)^2}$$

$$= \frac{(102.5 + 58.5)(102.5 - 58.5)(58.5 + 29)(58.5 - 29)}{(102.5)^2 (29)^2}$$

$$= 2.07$$

So, from Keele's data, we can easily get the right tuning frequency:

$$f_B = (2.07)^{0.32} (57)$$
$$= 72 \text{ Hz}.$$

And the cut-off frequency:

$$f_3 = \sqrt{2.07} \times (57)$$
$$= 82 \text{ Hz}.$$

From these calculations, we see that 13½% was not quite enough overvolume to make the enclosure precisely the right volume, but the difference is negligible. We will use this design. Construction details on this project appear in Chapter 15.

THE CLASSIC BASS REFLEX

Some audio workers have not yet accepted the Thiele data as thoroughly practical because, they say, there are too many variables in a ported box to predict by simple formulas how it should be built and tuned. Some of these critics of theoretical tuning go so far as to say that you should not go by any pre-determined tuning frequency: you should tune by ear. There is nothing wrong with tuning a system to your own taste, but unless you use some kind of guide you may end up with a boom box that offends every ear except the one that tuned it.

Regardless of the validity of the arguments for or against the practicality of the Thiele data, there are some speakers that it doesn't apply to, notably those with a high Q. Figure 7-1 shows that the highest Q listed for unequalized systems is 0.625. In the real world of speakers, with Q's ranging up to 3 or 4 times that high, what do you do? You may be tempted to put such speakers in a closed box, but they often have stiff cones that do not work in a closed box that is smaller than a piano. For such a speaker, you can try a classic bass reflex.

What is a classic reflex? Here are some of the descriptions that have been given for it:

1. Box is tuned to speaker's free air resonance.
2. Impedance peaks are equal in amplitude.
3. Impedance peaks are at equal distance from the free air resonance.
4. Impedance peak frequency ratios are equal.

At one time, there were other characteristics than these four points, such as making the port area equal to the effective piston area of the cone, but this practice makes such a large enclosure that those days are surely gone forever.

Figure 7-17A shows the kind of impedance curve you would get with a properly tuned classic bass reflex. The curve in Fig. 7-17B shows a box tuned too low, and Fig. 7-17C, a box tuned too high. When you compare the three curves, you can see that the frequency of the upper peak changes very little with the tuning; instead its frequency is determined by the size of the box. In the "B" curve, whose tuning is below the free air resonance, the maximum damping is low in frequency, which keeps the lower impedance peak under control. In "C," the maximum damping is higher in frequency than the free air resonance, so the upper peak is well damped. Before the Thiele data became known, some experts recommended a curve like that of "B" because it reduced ultra-low frequency distortion and controlled the cone at frequencies where excursion could get out of hand. Others recommended a "C" curve on the grounds that it damped the upper peak where boom was often more annoying than at the lower frequencies. The "tune by ear" advocate would claim that his method automatically chooses the best method for a given speaker.

Having looked at both sides of the "tune under" and "tune over" argument, let's consider how to tune to the free air resonance. The practice most recommended is given in Point 2 of a classic reflex: tune for equal impedance peaks (but this does not necessarily tell you the tuning frequency). For a given box frequency, tune until the lowest point in the valley between the peaks is at that frequency. Or apply any of the methods outlined in Test 7, Chapter 3.

What about Points 3 and 4? Upon a close examination, these points appear to contradict each other. Let's assume that the peaks should be equidistant in frequency from f_B: then in Fig. 7-17A, the lower peak, at 23 Hz, is 17 Hz below the tuned frequency of the box. That means that the upper peak should be at 57 Hz, 17 Hz above 40 Hz. Instead, the upper peak is at 69 Hz, well above the point predicted by Point 3. Although this equidistant theory has often been quoted, it does not work if the distance is measured in Hz unless the curve is plotted on log graph paper and the distance measured on the graph. This brings us to Point 4, the ratio of the peaks. Here the ratios are 69/40 and 40/23. These two ratios work out to about 1.73 and 1.74, another evidence of classical tuning, but evidence that sometimes must be abandoned if the peaks are made equal in height. In regard to the validity of Points 3 and 4, if Point 4 holds true, then Point 3 will also hold if the curve is graphed on log paper instead of common graph paper.

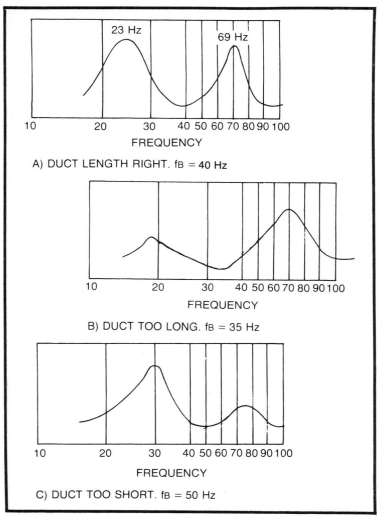

Fig. 7-17. Impedance curves for a speaker with a 40 Hz free air resonance in a classically tuned bass reflex.

Although Thiele says that making the impedance peaks equal in amplitude misses the point for his designs, some loudspeaker manufacturers, good ones, still say that this method gives a reliable way of tuning a bass reflex to a speaker. For any theoretical enclosure that does not work out, this is worth a try.

If you graph the impedance curve of a speaker in a ported box, there is one other ratio that can tell you something about the design:

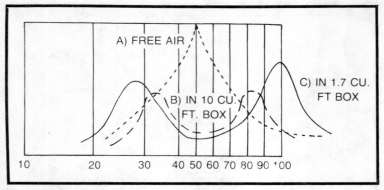

Fig. 7-18. Impedance curves of a speaker with an f_s of 50 Hz in free air (a), a 10 ft.³ box (b), and a 1.7 ft.³ box (c).

the ratio of the upper peak to the lower peak. In the early days of hi-fi, this ratio was kept to 2.4 or below. In the system shown in Fig. 7-17, the tuning is correct with an f_H/f_L ratio of 69/23, or 3. A box with a ratio of 2.4 would be much larger than one with a ratio of 3. Figure 7-18 shows the impedance curves for a speaker with a 50 Hz free air resonance when put into a 10 ft.³ box that is tuned to 50 Hz, and when put into a 1.7 ft.³ box tuned to 50 Hz. Notice that the smaller box appears to be more tightly coupled to the speaker, depressing the impedance more at f_B, but that the upper peak is at 100 Hz, about 4 times the frequency of the lower peak. This upper peak frequency can easily cause boom, and one of the dangers of a small box is that the frequency of the upper peak is high enough in frequency to sound oppressive.

Going back to Fig. 7-1, Keele's version of the Thiele data, the ratio of the peak frequencies, varies from about 2.4 to 7.6. In box design, a ratio of 2.4 gives a box with more than twice the V_{as} air volume, but at a ratio of 7.6, the box volume is less than a 1/10 the V_{as} volume. From the chart you can see the trend: speakers with low Q's are matched to small boxes (small at least by acoustical measurements) and speakers with high Q's are put into big boxes with low impedance peak ratios.

If you have a speaker with a relatively high Q, you may be tempted to choose an ultra low ratio and use an immense box. This could give the most extended bass response; however, as Keele has noted, this kind of combination gives high distortion. So which ratio do you choose for a speaker that does not fit the Thiele data?

The best answer is probably that given by Novak years ago: aim for a peak frequency ratio of 3, which you can get when V_B/V_{as} is equal to about 0.7. Add about 30% overvolume, and tune the box to the speaker's free air resonance, or for equal impedance peaks. For a speaker with a V_{as} of 5 ft.^{3}, you would need a box volume of 0.7×5 ft.^{3}, or 3.5 ft.^{3} plus 30%, about 4.6 ft.^{3}. If you get boom, put a damping pad over the speaker. Use Test 19 to check the damping. You will not have a genuine Thiele alignment, but for this kind of speaker and for many better ones, you should get good sound. Then experiment with the tuning if you have a good ear.

PROJECT 9: A UTILITY BASS REFLEX SPEAKER

This project came about because an acquaintance wanted some portable utility speakers to use with a reel-to-reel tape recorder that would give better reproduction than the built-in speakers. He had a strict limit on how much he wanted to spend, so he chose two 10 in. speakers with whizzer cones that cost $10 each. He also put a 2 ft.^{3} limit on the size of the box. He reasoned that the 10 in. speakers could take enough power to give good sound for small crowds in large rooms.

A test on one of the speakers, a Norelco 1065/M/8, showed these values:

$$f_s = 55 \text{ Hz}$$
$$V_{as} = 3 \text{ ft.}^3$$
$$Q = 1$$

The specification that stands out here is the Q, too high for satisfactory use with the Thiele alignment data. In fact, this is just the kind of speaker discussed in the previous section on the "Classic Bass Reflex."

Some quick calculations showed that by the 0.7 V_{as} method, this speaker should work satisfactorily in a box of slightly more than 2 ft.^{3}, or, with the overvolume, about 2.7 ft.^{3} The owner said no. He wanted a 2 ft.^{3} box and no larger, so he built a 2 ft.^{3} box (Fig. 7-19).

The original port measured 3 in. $\times$ 7½ in., and the duct length was 5¾ in. The first test showed the box frequency to be about 65 Hz, but we could not be sure of the exact point. Examination showed that the builder, who had never before made a speaker box, had permitted some leaks. We stopped those with silicone rubber and added some fiberglass. The new frequency was 58 Hz. We reduced the port area by gluing pieces of ¼ in. plywood inside the duct walls, which brought the box frequency down to 52 Hz. At this point, we read the three critical frequencies at 93: 52: 29. These show that the ratio of f_H/f_L was about 3.2, which was higher than we wanted, even though we had put extra fiberglass in the box. When we compared upper peak frequency to box frequency, and box frequency to lower peak frequency, the ratios were equal, indicating correct tuning, but the upper peak was higher in amplitude, indicating that the tuning was too low. By one guide the box was tuned, by the other, it was not.

Fig. 7-19. Project 9 with grille frame removed and before final tuning duct was added.

We shortened the duct to 4 in., trying to make the peaks equal. We overdid that: the lower peak was now higher than the upper one, with critical frequencies of 99: 63: 30. We considered stopping here, but then we listened to some music and went back to work. Shortening the duct should have been all right by Thiele's design methods, but an unexpected gremlin had entered the box. There was a sudden increase in mid-range response that gave an unnaturally bright quality to the sound. We traced the trouble to the sloping speaker board and the shortened duct; the angle of the backwave was just right to reflect through the shorter duct. We installed the longer duct again, enlarging the area somewhat. The final ratio of frequencies was 97: 59: 30, and the peaks were about equal in height. We stapled a damping pad over the speaker. At the final listening test, the owner was delighted with the sound.

The experiences with this speaker show that unexpected complications can arise from minor changes in an enclosure. They also suggest that anyone who designs and builds a speaker system by theory alone can end up with poor sound. After all, there are a lot of variables.

Construction details for this project appear in Chapter 15.

8

Labyrinths and Transmission Lines

Several loudspeaker manufacturers have recently converted some of their production to transmission lines, which suggests that the large lines must have virtue—or at least appeal. The people who make them claim the virtue of non-resonance. A pipe, or *line*, is actually an inherently resonant device, but the damping material in the long line depresses the resonances of the line. Unfortunately, this damping material also cuts the efficiency. While closed box speakers are small and usually inefficient, transmission line speakers have the double disadvantage of being large and inefficient.

In most transmission lines, the lowest frequency sound is not absorbed. Instead, it emerges to reinforce cone radiation down to 20 or 30 Hz, or to whatever frequency that phase shift produces cancellation. The frequency of the cut-off point by phase shift depends on the length of the line.

The damping material in the line acts somewhat like a shunt capacitance in an electrical circuit: the more damping material used, the lower the frequency of attenuation. This corresponds to increasing the value of the shunt capacitance in the electrical circuit.

Transmission line enclosures have attracted considerable attention during the last decade. The cause of much of this attention has been the element of mystery about these maze-like enclosures. Another reason for the appeal of transmission lines was the undeveloped science of matching speakers to the simpler-to-build reflex boxes. Now that the design of reflex enclosures has been more accurately analyzed by Thiele, Small, and others, the transmission line will face the increasingly difficult challenge of proving its worth aganst the less expensive, easier to build, and more compact ported box. But even if experimentation should prove that a more compact reflex produces a flatter, more extended bass response, that would not necessarily mean the end of the transmission line. There would surely be people who will choose a certain type of enclosure because they like it on principle, because they think the quality of the bass is better than that of other types, or maybe because they are different.

HOW TO DESIGN A QUARTER WAVELENGTH LABYRINTH

Measure the driver's free air resonance, according to Test 1, Chapter 3. Calculate the tube length that will produce an effective tube length of one-quarter wavelength at resonance by:

$$1 \text{ (in ft.)} = \frac{c}{4 f_s} - 1.7 r$$

where:
   1 = length of pipe
   c = speed of sound in air, about 1120 ft./sec.
   r = radius of pipe in ft.

If the pipe has a rectangular cross section, you can find r by:

$$r = \sqrt{A/\pi}$$

The correction from a full quarter wavelength line accounts for the small amount of room air that vibrates with the air in the line, making the line act longer than its physical length.

This formula is all right for a straight pipe with no damping materials. Not only would such a pipe be rather long, but because of numerous resonances, it would sound terrible. When the line is folded, the correction factor should probably be increased.

HOW TO TUNE A LABYRINTH

Figure 8-1A shows that the double humped impedance curve of a speaker in a quarter wavelength straight pipe is similar to that of a reflex enclosure. Like a reflex, the quarter wavelength labyrinth provides maximum damping for the speaker at resonance. Most practical labyrinths would show less damping at resonance than the deep valley in Fig. 8-1A because they normally have damping materials on the walls. This curve was run with an empty pipe to illustrate the damping effect.

To tune the labyrinth, install damping material on the inside surfaces, then adjust the length until the valley of the impedance curve occurs at the frequency of the free air resonance. Notice that you should add damping material first, so that any change that it makes on effective length will be automatically compensated for in the tuning.

The practical way to get the right length is to make the length adjustable. In Fig. 8-2D, the labyrinth can be shortened by moving the port to the left, as viewed from the front of the cabinet.

To save time and money, you can build a temporary labyrinth out of celotex (Fig. 8-3). If you do not glue the parts together, you can remove sections to make major changes in its length, and test again until the impedance curve looks right.

A question that often arises is that if labyrinth action is similar to that of the reflex, why not build a reflex enclosure instead of the larger and more expensive labyrinth. A good question. To that argument we can add that the long pipe of the labyrinth produces peaks at even multiples of a quarter wavelength: for example, at a half wavelength and at a one wavelength.

TRANSMISSION LINES

If a pipe is stuffed with damping material, the impedance curve changes shape. Fig. 8-1B shows what happened to the impedance curve of the speaker in the quarter wavelength pipe when the pipe was stuffed. The peak in the curve occurs at about the free air resonance of the speaker, but the curve is much flatter than that of a speaker in free air or in a simple box. Transmission line advocates make much of this flat impedance curve, but it should be remembered that an impedance curve does not represent sound output for modern amplifiers with low internal resistance.

The pipe of Fig. 8-1B is too short, and the woofer's free air resonance is too high for an ideal transmission line, but the stuffing in the pipe makes a flatter curve regardless of pipe length. If the pipe were longer, the bump in the curve would be moved to the left on the frequency scale, particularly if a lower resonance woofer were used. For some transmission lines the rise in impedance is very gradual and occurs at the low end of the frequency reproduction band.

If you build a transmission line enclsoure, you do not have to worry much about critical tuning. The typical length of a line is seven feet, but it will require critical stuffing, enough damping material to damp reflected mid-range sound and to kill line resonances without choking off the lower bass. I.M.F., a company headed by I.M. Fried that makes several transmission line speaker systems, talks about achieving a correct "free flow" cross sectional area in the tube, meaning the area of the line that is above that occupied by the fibers of the stuffing material.

One approximation is to make the tube area equal to the effective piston area of the driver. Generally, when this rule is applied, it

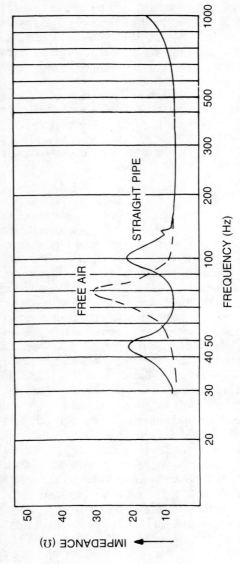

A) IMPEDANCE CURVES OF SPEAKER IN FREE AIR AND IN A QUARTER WAVELENGTH STRAIGHT PIPE.

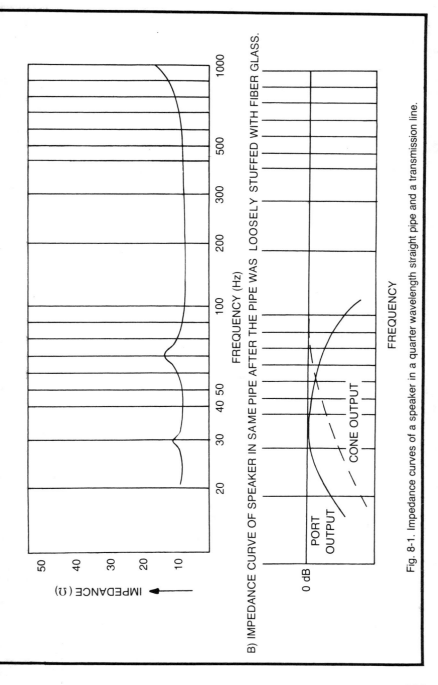

Fig. 8-1. Impedance curves of a speaker in a quarter wavelength straight pipe and a transmission line.

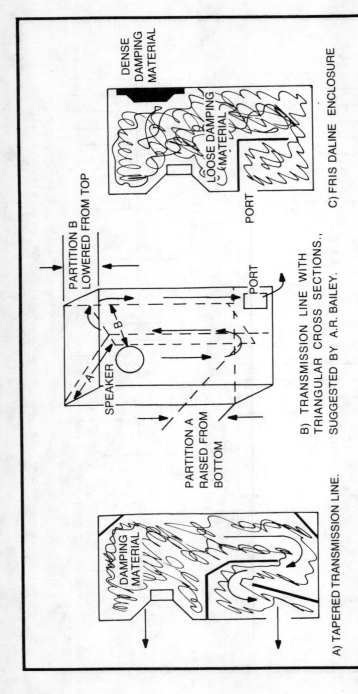

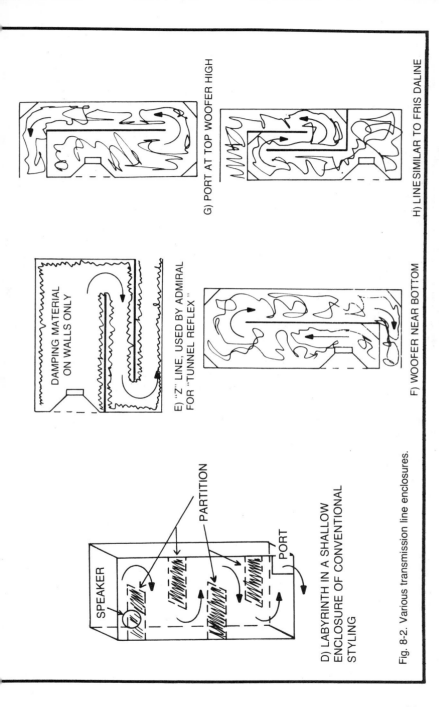

Fig. 8-2. Various transmission line enclosures.

is considered to find a minimum area: the section of the tube near the driver is usually much larger, having a cross-sectional area of at least 1.5 to 2 times that of the driver. Some transmission lines have been designed with a minimum area near the port that is about 50% that of the driver. This is another piece of evidence that at this stage, the practice of transmission line design is more of an art than a science. One difference between the approach to transmission lines and the approach to closed box or reflex enclosures is in the kind of specifications mentioned for the drivers. While one can predict with considerable precision how a driver with a certain resonance frequency, Q, and V_{as} will behave with a certain box volume and port tuning, there is little basis for similar predictions for transmission line drivers. Some experimenters have recommended low Q woofers for lines, while others have mentioned that a low Q woofer is easily overdamped in a stuffed line, so they prefer high Q sheakers.

TRANSMISSION LINE GUIDES

Lacking the more precise design rules of other enclosures, here are some suggestions for making a transmission line system:

1. Choose a woofer. Give preference to woofers with a butyl rubber or foam roll surround. There seems to be a rough concensus on the 10 in. size woofer or its equivalent in piston area among several manufacturers, some of them English, that make line systems. Larger woofers require an enormous enclosure, and smaller ones have reduced bass handling power. The only difference in designing a line for a different size woofer is in scaling the cross-sectional area to match that of the driver's effective cone area.

2. Make the minimum cross-sectional area of the pipe equal to the piston area of the driver. Make allowance for the area occupied by the damping material in the line. To get an estimate of the area occupied by the solid fibers of the material, compress a sample of the material between your thumb and forefinger and compare the compressed thickness to the expanded thickness. For example, if you can compress a 1 in. thickness of fiberglass to a thin sheet that is about 0.1 in. thick, you can consider that the fiberglass will occupy about 10% of the tube area. In that case, you would add 10% of the area occupied by damping material to the total line area to allow space for the damping material.

3. Make the area of the pipe near the driver about twice that of the effective piston area of the driver. Taper the line away from the driver so that the minimum area occurs at the port end.

4. Use reflectors at the bends in the line near the driver. These reflectors should be angled at 45 degrees to the sound path.

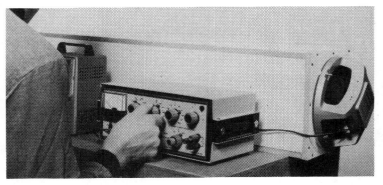

Fig. 8-3. For transmission line experiments you can make a temporary pipe out of celotex, but a folded pipe of the same length will act longer.

    5. Use standard, approved construction techniques to make a rigid, non-resonant enclosure.

    6. Use some kind of support for any kind of stuffing that settles with time. Large mesh cheese cloth or fish netting will do this job. Long-fiber wool needs support more than other materials and should be moth-proofed. Dacron is light and has little tendency to settle. It seems to be the second choice among transmission line builders. Fiberglass is more dense, but its density varies, and it must be used with care.

    7. Make one removable panel, or more as necessary, so that you can stuff the line and make adjustments in the stuffing.

    8. Test the system with various amounts of stuffing by:

        a. Impedance test 2, Chapter 3. Try to get the smoothest, flattest curve possible without overdamping.

        b. Oscilloscope test, set-up B, Chapter 3. After stuffing the line, check for port output by covering the port while you observe the test pattern on the scope. Any change in the pattern when you close the port indicates phase shift and that the port is radiating at that frequency. No change indicates overstuffing, unless you want total absorption in the line.

        c. Frequency response test 9, Chapter 3. Use the near-field microphone technique of Method II to monitor the output of the cone and the port. Because of the low frequencies involved, you will need a good microphone for this test to mean anything. The ideal response is shown in Fig. 8-3, with the enclosure acting as a crossover device between cone radiation and port radiation.

    9. Expect to spend considerable time experimenting with damping material.

    10. If you run into trouble, use the line as a dog house and design either a closed box or reflex enclosure for your woofer.

9

Omnidirectional Speakers

Omnidirectional speakers produce the kind of sound difference that you can hear immediately. There is no mistaking the immense spaciousness of a good omnidirectional system that gives an illusion of three-dimensional depth and spread, expanding the apparent size of the source, making it appear big. Some listeners say it sounds bigger than life. In fact, some experts complain that omni-directional speakers make every vocalist sound like Mr. or Ms. Five by Five. Or wider. The omni lovers admit that this sometimes happens, but, they say, we have smashed the wall that used to separate the performers in a recording from the listener. Now, instead of hearing them through a hole in that wall, they are spread out before you or you are in the middle of them. Most listeners seem to like the effect and consider omni-directionality an added dimension in life like sound.

In addition to the objections to wide field vocalists or solo instrumentalists, there is another line of criticism aimed at omni speakers. Some people say they are good for mono recordings because they add a pseudo-stereo effect, but that they degrade stereo sound by producing such a diffused image that it is impossible to precisely locate instruments in space. Omni fans reply that it is not necessary to be able to pinpoint the placement of a single instrument to have good stereo reproduction. In fact, they note, you can not do that at a live performance.

And so the argument about omnidirectional speaker systems goes on. You do not have to rely on expert opinion to decide whether you should have an omnidirectional speaker. As mentioned in the introduction to this book, experts can furnish facts and give advice, but only you can decide what kind of speaker system you want. The best way to make the final decision on any speaker is to listen to it. Just make sure you listen long enough to eliminate the novelty effect of hearing something new, and to find out how well the sound wears after a long listening session.

Omni-directional speakers produce their special effect by producing a high ratio of reflected to direct sound. In a true omni system, which radiates over a full circle, the sound goes out from the enclosure like the waves do when you drop a rock into a quiet pond. As the sound hits the room walls and other objects, it is reflected around the listening area (Fig. 9-1). With a good omni system you may notice very little change in sound character as you move around the room because most of the sound has been reflected around the room before you hear it. Direct radiator speakers produce less reflected sound, and the source of the sound can be more easily located by the listener. As you approach a direct radiator speaker, the sound intensity increases; but an omni-speaker produces sound that has little apparent variation in intensity.

KINDS OF OMNI-DIRECTIONAL SPEAKERS

There are many ways to make an omnidirectional speaker. The simplest is to put a single wide range speaker in the top of an

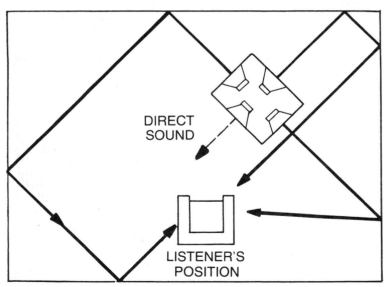

Fig. 9-1. How an omnidirectional speaker system produces a high percentage of reflected sound. Changing the angle of the speaker cabinet can modify the sound at any given position.

enclosure, facing upward into a reflector. There is a possibility of coloration from this method, but it does widen the source and it is simple and inexpensive. Some commercial systems have used upward facing speakers with reflectors. There is one caution to observe with this kind of system: do not use a speaker with a heavy cone. The constant force of gravity on a heavy cone may cause it to drift lower on its suspension, which would put the voice coil in a region of non-uniform magnetic field. A useful rule of thumb is to limit the diameter of upward firing speakers to 8 in.

Another way of producing an omnidirectional speaker is to use a multiple array of full range speakers, mounted so they can fire in all directions. Speakers in an array help each other by increasing the radiation resistance so they can produce better bass response, particularly if more than one speaker is mounted on a side. This kind of omni system is inexpensive; there are no crossover components, and several small speakers can often be purchased for less than enough high quality woofers and tweeters to make an omnidirectional woofer-tweeter system. The array will have a more limited frequency response unless special tone control circuits are available that boost the low bass and high treble. Unfortunately, such tone control circuits make large power demands on the amplifier.

Most high quality commercial omni speakers have crossover networks that divide the sound between low and high frequency speakers. Typically one or two woofers handle the naturally omnidirectional low frequencies with at least two or more tweeters for the highs. The tweeters usually fire out of opposite sides if there are two, and out of three or four sides if there are more. An added dimension can be obtained by mounting one tweeter so that it fires upward. The separate woofer-tweeter arrangement is the surest way to get wide range omni-directional sound.

PROJECT 10: AN OMNIDIRECTIONAL SPEAKER SYSTEM

When an omnidirectional speaker has a single woofer, it is desirable to use tweeters that can reproduce sound down into, or even below, the mid-range frequencies. To keep the price low, the tweeters must be somewhat larger than the typical conventional tweeter so they can handle considerable power at fairly low frequencies. A compensating feature is the high percentage of reflected sound that distributes highs over the listening area more effectively than would a single tweeter of the same size.

The omnidirectional system shown in Fig 9-2 was developed for good appearance as well as for decent sound. A pair of them can serve as lamp tables at the ends of a sofa or other long piece of furniture. The enclosure is compartmentalized, the woofer in the

Fig. 9-2. Project 10, an omnidirectional speaker system used as a lamp table.

lower compartment and the mid-range tweeters in the upper. The system was planned and built to use five drivers, a single woofer, and four tweeters, but final tests with a super tweeter showed improved performance with it. The problem of how to add a super tweeter without going to a complicated crossover network was solved by adding a Motorola piezo-electric tweeter. This tweeter requires no network at all because its impedance rises at low frequencies, chopping off the lows from its branch of the circuit. Piezo-electric tweeters have a wide frequency range with good transient response. In this application, the super tweeter fires out the back of the enclosure so that all of the upper highs are reflected from the wall behind the speaker. This trick widened the apparent high frequency source considerably over that obtained by the four mid-range tweet-

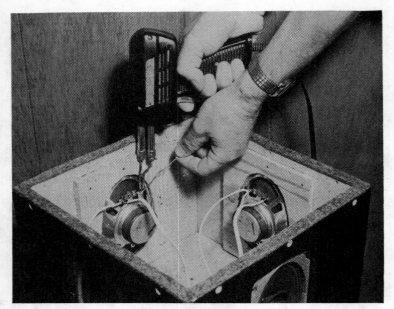

Fig. 9-3. Project 10. Wire opposite tweeters together out-of-phase.

ers, and also balanced the output of the super tweeter to the other drivers without needing a control for the super tweeter. With the six drivers in each enclosure, a pair of these speakers gives a good omni-directional spread with a soft airiness from the super tweeter adding a live quality to the music, but this effect was not achieved without some difficulty.

For one thing, the box has a square cross-section, which is good for appearance in this tower-shaped enclosure, but not the best shape for opposing drivers. An earlier omni-directional system using the same Norelco woofer and four Norelco 5 in. full range speakers worked out very well. That system used a triangular column that was compartmentalized like this one, but three of the tweeters fired out from the three sided enclosure, while the fourth one faced upward. The low priced Norelco 5 in. speakers used in that system are no longer available, so Norelco 4 in. × 6 in. speakers are used in this one.

When this system was first set up, the four oval speakers were wired in phase, with all the cones pushing outward together and inward together. That set-up raised the speakers' resonance from 135 Hz in free air to about 200 Hz in the enclosure. This resonance was more pronounced than the resonance in the earlier triangular box, and the peak there coupled with higher frequency peaks caused by the standing waves in the square enclosure caused uneven response unless the crossover frequency was raised into the midrange band. In order to keep the crossover frequency down, the wiring of the oval speakers was changed from all in-phase to opposite speakers out-of-phase (Fig. 9-3). This lowered the resonance to 135 Hz, the same frequency as in free air. The opposite speakers being now wired out-of-phase, they moved in the same direction with respect to the compass when a signal was applied, which maintains a more constant pressure in the tweeter compartment. If the oval speakers had been carrying the bass, such a situation might have caused some bass loss, but that is of no consequence here. A falling bass response is desirable for a tweeter or mid-range speaker anyway. The out-of-phase upper speakers are fed through a 12 μf capacitor. The value of the capacitor was selected by test, not by calculations or crossover charts. The problems of the square box

Fig. 9-4. Project 10. Screws through the sides hold down the top. The super tweeter is in the rear.

were solved by filling the upper compartment with loose polyester batting. In this particular enclosure, loose polyester batting gave better sound than the typically more dense fiberglass, but there was little difference between the polyester and the loose house insulation grade fiberglass.

One way to improve this system is to use a larger enclosure with a 10 in. woofer, but that would seriously alter the appearance. The Norelco 8065/ W/8 woofer that is used here is a good one, but you may have trouble finding this particular model because the company has just recently introduced a new model, the 8066. I have not tested the new model myself, but I have heard reports that it has a higher frequency of resonance and so a higher bass cut-off than the 8065. Of course, any good 8 in. high compliance woofer can be substituted for the one shown here, but you will have to make the cut-outs match the speaker. If the woofer you choose has a low Q, you can use a vented enclosure and gain some bass range over the sealed box chosen for the Norelco woofer. The closed box shown in Fig 9-4 is less critical in the kind of speaker that can be used, of course. If you use a woofer with a high Q, you should use more fiberglass or Dacron batting in the woofer compartment than is specified in the plans shown in Chapter 15.

10
Multiple Speaker Arrays

In the early 1960's, there was an epidemic of multiple speaker arrays, small speakers wired together without a crossover network. Some people who already had expensive speakers built the arrays and claimed that they sounded as good as their big name models. Then some kill-joy engineers ran tests on the "Sweet Sixteens" and other similar arrays which showed that their bass range was limited and that their mid-range was peaky. When such hard facts did not completely kill the concept, it became apparent that multiple speakers have a great appeal and that when they are used in the right way, there may be some good reasons for their popularity.

The theory behind the multiple speaker array is that the power is divided among many speakers, each one handling a small fraction of the total. Any speaker has lower distortion and a flatter response at a power lower than its rated power, so the theory is that small speaker arrays should therefore have low distortion and smooth frequency response.

There is more to the story than that, both pro and con. When two speakers are mounted close together, each cone helps to load the other cone, increasing the radiation resistance. Because of this mutual loading effect, two identical speakers close together can theoretically radiate four times the power at low frequencies that a single speaker can. If small speakers with light cones are used, their transient response will be better than that of a large speaker.

So far, so good, but there are some flaws too. Although bass loading occurs and somewhat reduces the frequency of resonance, low frequency response will roll off below that frequency. The smaller and cheaper the speaker, the higher the frequency of bass cut-off will be. If the speakers are installed in a closed box, the air trapped in the box will raise the frequency of resonance still farther. You could get lower bass by removing the back of the box, but that would remove some of the restoring force on the cones and increase bass distortion. Of course, you could use enough speakers to limit distortion, but the system would then be unwieldy and expensive.

One solution would be to use better speakers, but again such a system could become as expensive as a conventional one. One way to avoid this predicament is to use a multiple array for mid-range only, using separate woofers and tweeters. Even then the concept must be used with some care, or it will produce a system with poor sound dispersion.

MOUNTING PATTERN

Instructions for mounting multiple speakers nearly always show them installed in a rectangular pattern on a common baffle with a caution that they should not be put in a straight line. Mass mounting on a large baffle does improve coupling by putting many drivers closer together; it also widens the source, reduces horizontal dispersion, and produces peaks and dips in the frequency response by introducing different path lengths from the speakers to the ear of the listener. If the speakers are used for mid-range duty, line mounting is much better because the total width of the line source is that of a single speaker, and if the array is to be used purely for mid-range, it does not need the additional low frequency loading provided by massed mounting.

Another way to use many small speakers without poor dispersion is to use an angled baffle. If the front of the enclosure as seen from the top is convex, the speakers will face out over the room and better cover the listening area. At the higher frequencies, the sound from adjacent speakers which are firing at different angles will not overlap as much as on a flat baffle, and the dips and peaks will be lsss evident.

When multiple woofers are used in an array, they should be massed with close mounting in a square pattern to take the greatest advantage of mutual coupling. The crossover frequency should be placed low for this kind of system. Ideally, it should be low enough for the highest frequency produced by the array to have its wavelength be about equal to the width of the array.

WIRING CIRCUITS

At its best, an array can give good sound. Even though the speakers have similar individual resonances, their resonances will shift in the array because the different speakers in the array see different degrees of mutual coupling (Fig. 10-1). The shift may not be great, but it does broaden the resonance, which is always a desirable goal.

Another way to smooth the output at resonance is to use speakers with widely spaced resonances and wire them in parallel. There is, of course, a limit to how many speakers can be wired in parallel; if the impedance is too low, the output stage of the amplifier will draw excessive current and get hot, or even suffer damage. This method can still be employed if you use enough of each kind of speaker, put one of each kind in each parallel branch of the circuit as shown in Fig. 10-2, and then wire the several branches in series to make up the right total impedance. An easy way to plan such a system is to let n equal the number of speakers in the array. Then you should use ($\sqrt{n}$) speakers in each parallel circuit with ($\sqrt{n}$) branches. For example, if you want to use 16 speakers of 8Ω impedance, you would use 4 ($\sqrt{16}$) speakers in each parallel branch. One branch would have an impedance of 2Ω (8/4), so 4 ($\sqrt{16}$) branches in series would bring the impedance back up to 8Ω. Obviously you must use a number for n that has a whole number for a square root.

Because arrays can have an efficiency that is higher or lower than the other drivers, you may occasionally want to make the total impedance of a mid-range array slightly higher or lower than the rated impedance of a woofer or tweeter. In this case, instead of using the square root method of determining speaker numbers, you would divide the number of speakers in the array by a small whole number and use the quotient as the number of branches. Then add up the impedances and see if the total impedance is what you want. For example, a 6 speaker array can be divided by either 3 or 2. When we divide by 3, giving 2 branches, the number of speakers in each branch is 3. If they are 8Ω speakers, the impedance of each branch is 8/3, and the impedance of the 2 branches in series is 16/3 or about 5-1/3Ω. That is all right if the array is less efficient than the other speakers in the system, but suppose it is more efficient and we want to limit its output without going to the expense of an L-pad. Then we would reverse the numbers, making 3 branches of 2 speakers each. The impedance would be 8/2 or 4 Ω, and the impedance of the array would be 12Ω. The wiring system you choose will depend on the impedance of the speakers as well as on their efficiency. For example, some small speakers might have an impedance of 5 or 6 Ω, others up to 9 Ω or higher. With the low impedance speakers you would undoubtedly want to use the 3 branch system, but for a higher-than-normal impedance speaker you might find that the 2

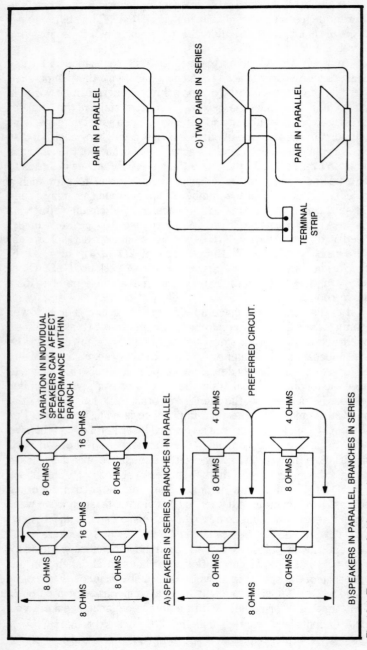

Fig. 10-1. Two ways of wiring speakers in arrays so that the circuit impedance is equal to that of a single speaker. Theoretically there should be no difference in performance between (a) and (b). However, because of speaker variation, (b) is preferred, and (c) is the pictorial diagram of how to wire such an array, copied from a JBL bulletin.

branch circuit with its lower total impedance would balance the system better.

If you are building a full range array, you may want to use an n that has no whole number square root. Say you have 20 speakers on hand and want to use all of them. That's all right, but the total impedance will not equal the rated impedance of the speakers. Try to use a wiring arrangement that will put the final impedance between 5 or 6 and 16 Ω. Just make sure you have an equal number of speakers in each branch and identical branches in the circuit so that each speaker will handle an equal amount of power. Some experi-

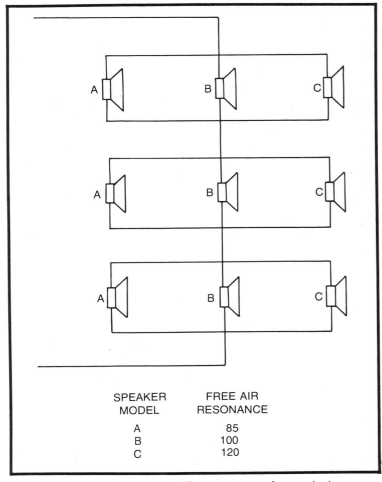

Fig. 10-2. How to use speakers with different resonance frequencies in an array.

menters forget to do this and have built arrays in which one or two speakers were handling up to half the total power.

ENCLOSURES FOR ARRAYS

It is possible to get doubled radiation from an array by installing the speakers in a shallow open back enclosure. The radiation pattern as seen from above is then a figure eight. This kind of forward-backward radiation can produce very realistic sound if positioned right. After all, most musical instruments radiate sound in all directions. However, this radiation is critical as to room placement; if the speakers are placed close to a wall surface with the speaker board parallel to the wall, reflections will color the sound by adding peaks at frequencies where the reflected waves are in phase with the speaker cone. This limits the use of dipole radiators to situations where there is considerable leeway in cabinet placement, and to people who do not mind spending considerable time in finding the best position for the speakers.

When an open back enclosure is used, it should be shallow sided, usually not more than about 3 in. or 4 in. from the rear of the speaker board to the back of the sides. A deep, open backed enclosure can produce a nasty upper bass resonance. One example of this kind of reproduction was found in the early-day console radios that usually had open backed cabinets of just about the right volume to produce a resonance at approximately 200 Hz. While this gave the illusion of a heavy bass response, a comparison with a high fidelity system showed that the open back cabinet produced most of its bass in a narrow band of frequencies and was a "one-note" bass. Keeping the open back enclosure shallow reduces this kind of bad resonance.

LINE ARRAYS FOR PA SYSTEMS

Several manufacturers make commercial line arrays for PA systems and with good reason. These systems are often used in such bad acoustical environments as gymnasiums where the room is too reverberant. A vertical line array will produce sound that has good horizontal dispersion, but limited vertical dispersion. That's good. Vertical dispersion only sprays sound onto the floor and ceiling, adding to the multiple reflections already present in the too-live room and contributing to the jumble and confusion that is typical of these places. By careful placement, a set of line arrays can distribute direct sound to a crowd with little of the overlapping between widely spaced arrays that would otherwise confuse the sound image. A higher percentage of direct sound to the listener makes for greater clarity in a live room where the normal reverberation time is too long.

Some commercial line arrays place the speakers in a concave line so that vertical dispersion is even more limited than in the standard straight line. Another refinement is to taper the response of the array from a full frequency range in the center to a progressive roll-off of highs for the speakers away from the center. For example, if 9 speakers were used in the array, the center 3 would have no electrical frequency compensation, but the upper 3 and lower 3 might have chokes in their feed line to limit their high frequency response. At the center branch, the middle speaker would have no acoustical treatment, but the speakers above and below it might have a thin sheet of fiberglass or other material to roll off their highs. This is a hypothetical system, but both electrical and acoustical filters have been used to control the high frequency response of all but the center speaker. By reducing the diameter of the source as the frequency increases, the filters prevent the overlapping radiation patterns at high frequencies that can produce lobes in high frequency coverage because of interference. Such lobes can give the effect of a changing sound quality as you move around the room and pass through them, or the effect of poor highs if you are sitting in the

Fig. 10-3. A square mounting pattern like this is good for mutual bass loading, but bad for mid-range response and high frequency dispersion. This nine-speaker system gave poorer sound than four of them had in a smaller box.

wrong position. Another, simpler way to solve the problem is to use a separate tweeter system for the high frequency band.

ROCK CONCERT LINE ARRAYS

When, as for rock concerts, high power ability is needed in extremely large rooms or for outdoor crowds, larger speakers are used in the arrays. Like musical instrument speakers, the drivers for these high power speakers may have a lower compliance than would a high fidelity speaker made for home use. The stiffer suspension permits the speaker to absorb more power without damage, and an ultra low frequency response is seldom needed for rock music where middle bass is more important. Another critical consideration for such speakers is the kind of voice coil cement. Special cements, such as epoxy cements, must be used to prevent the voice coil from coming unstuck during long sessions of high power use. An experiment conducted by Cleveland Electronics, the makers of Cletron loudspeakers, showed that after four hours of operation at a 200W power level, a 12 in. guitar speaker with a 3 in. voice coil had an average voice coil temperature of 518 degrees of Fahrenheit, or 270 degrees Celsius. Because of the heat, the resistance of the voice coil had doubled, which restricted power to the speaker. This data indicates a big advantage of multiple speakers for such high level power applications.

PROJECT 11: A MID-RANGE ARRAY

In this system, six 4 in. × 6 in. speakers were used for the mid-range with a tweeter at the top of the array and a woofer in a separate enclosure. Almost any woofer will do if part of the project can be designed to suit the space for the woofer. One possibility is to put two woofers in a single box with separate terminals on the outside of the box so that one woofer can be wired to one array, and the other to the array in the second channel. This set-up would occupy less floor space than conventional systems (Fig. 10-4), although separate woofers can be used with the array mounted on top of its woofer enclosure, unless the woofer enclosure is too tall.

Ideally, the mid-range speakers should be in a single line, but that would make a very tall cabinet that would need support. No acoustical treatment need be used in the back of the enclosure unless wall reflections are a problem, and then you should cover the back of the speakers with a blanket of fiberglass.

This speaker has some unevenness of response in the mid-range, but there is a quality to the sound that is different from that of speakers installed in a closed box. Considering its low price, the sound is impressive.

Fig. 10-4. Project 11. A mid-range column mounted on top of a small woofer enclosure which contains a high compliance side-firing 6 in. × 9 in. woofer in a ported box.

PROJECT 12: SINGLE COLUMN STEREO SYSTEM

Unlike Project 11, which is versatile and has many applications wherever a low priced stereo speaker system is needed, this one was designed for a special purpose. It is somewhat similar to one advocated by Norman Crowhurst many years ago for noisy basement rooms where space is at a premium. It includes a complete double speaker system for single enclosure stereo. Because it is a tall, narrow column, and occupies only about a square foot of floor surface, it saves valuable space (Fig. 10-5). With both channels coming out of one column, the stereo image is somewhat compressed, but some people who have heard it find this more natural than divided speaker stereo.

Fig. 10-5. Project 12. A single column stereo system is a real space saver.

The 6 in. × 9 in. speakers face out three sides of the column, 4 speakers to a side. In its normal, forward position, the front 4 speakers are divided: speakers 1 and 3 (numbered from the top) are connected in the left channel circuit, and speakers 2 and 4 are connected in the right channel circuit. With the "front," mixed side forward, the stereo image is most restrained, but when the cabinet is reversed so that side faces away from the listener, the image is broadened.

Three models of 6 in. × 9 in. speakers were used in the system, 4 speakers of each model. The resonance frequencies of the three kinds of speakers varied from about 55 Hz for the model with the lowest frequency to about 75 Hz for the model with the highest frequency, the third model having a resonance slightly lower than 75 Hz. The speakers were installed in the enclosure so that they could be wired in branches of three speakers, each branch containing three dissimilar speakers, one of each model. Two branches were wired in series per channel, making an impedance of about 5 Ω per channel. With no back on the enclosure, but with the two compartments filled with fiberglass, each system resonance is about 65 Hz.

This system gives a good account of itself for the small amount of floor space it occupies. Construction is simple and straightforward, with no unusual or critical features. It is a good choice for special situations such as a recreation room or anywhere that a single speaker is preferred for stereo program material. Its efficiency is high enough that it can be driven by a modest amplifier.

PROJECT 13: A SMALL PA COLUMN

This column speaker uses a special driver designed for speech distribution systems. It has a frequency curve tailored for better voice intelligibility.

When the cabinet is filled with fiberglass, the system resonance occurs at 170 Hz with a gradual roll-off below that point. While this system is adequate for some kinds of music, it is primarily designed for good horizontal coverage for the spoken voice.

Plans for these three array projects are presented in Chapter 15.

Unusual Speaker Enclosures

Over the years people have tried to mount speakers in many different kinds of readymade enclosures to save the time and expense required to build one. Some of these substitute enclosures work the way most makeshift things do—poorly. Others are quite acceptable, especially if some care is used to see that no rules for good enclosure construction are violated. Here are a few that serve a useful purpose as test speakers or book ends and give good performance for their cost.

QUICK TEST SPEAKER

For troubleshooting work, it usually is not necessary to have a wide range, high quality speaker; any working speaker will show whether a defect in a piece of audio equipment is in the speaker or in some other part of the circuit. If you need an extra speaker for this purpose, you can make one in a hurry with a 4 in. speaker and a plastic milk carton.

Make sure the carton is clean and dry. Position the carton so that the carrying handle is near the back and cut a hole for the speaker in the front. Drill or punch holes in the carton for speaker mounting bolt holes and for two banana jacks in the lower front side of the enclosure. To drill the holes, you must support the thin plastic with a small board. Solder internal leads to the banana jacks and mount the jacks in the prepared holes. Damp the carton by stuffing damping material through the speaker hole until the entire volume is loosely filled. Place bolts in the mounting holes so that their heads are inside the carton. Front mount the speaker. When tightening the nuts on the mounting bolts, you may have to grasp the threaded bolt end with a pair of pliers to keep the bolt from turning as you apply force to the nut with a small wrench. Prepare some test leads with banana jacks on the speaker end, and alligator clips or other appropriate connectors on the other end. This completes your test speaker. It's not much to listen to, but it's very convenient with its built-in carrying handle.

BOOK END SPEAKERS

Many speakers are called bookshelf speakers, but most of them are too big to fit into a bookshelf. Miniature speakers can serve double duty as bookends if you glue a felt or thin foam pad on the bottom and put something on or in them to give them stability to keep them from moving away when the books lean against them. The simplest trick is to fasten a sheet of smooth, thin, but stiff sheet metal to the bottom of each speaker box allowing the sheet metal to extend about 4 in. to 6 in. from the edge of the box so that at least a couple of the books will sit on the metal. The metal must be smooth, or it will scratch furniture or snag the books.

Another way to add to the speaker's inertia is to load the cabinet with a heavy weight. Make the box slightly larger than necessary and glue a brick, or a couple of patio tiles, or almost any heavy material, to the upper surface of the bottom panel (Fig. 11-1). Or you can mount the weight with screws and washers. If you have any cement or ready-mix concrete available, you can drive several nails into the bottom panel for anchor pins and pour a layer of concrete into the bottom of the enclosure. If the back is glued on, you can put the concrete in through the speaker hole. If the back enclosure is to be removable, make a temporary form as high as you intend to pour the concrete, and paint the front surface of it with oil so that the concrete will not stick to it. When the concrete has set, you can pull out the

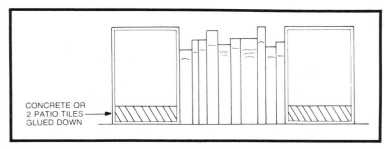

Fig. 11-1. How to load small speaker boxes with ceramic patio tiles for bookend speakers.

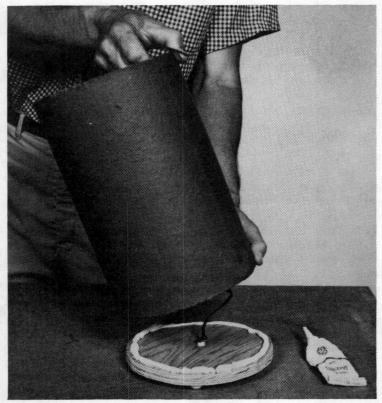

Fig. 11-2. Gluing a 12 in. section of 6 in. (I.D.) tile to its base. This size enclosure is good for 5 in. high compliance speakers.

form and install the back. These weighted enclosures are good to put any place where a lighter box might be easily pushed out of position or off its support.

CERAMIC TILE ENCLOSURES

Ceramic tile is used for drain pipes, which are round pipes, and for flue liners, which are rectangular cross-sections. In the early 1960's experimenters began using 4 ft. sections of drain pipe for columnar speaker enclosures, mounting the speaker in the top, facing upward. These had some pipe resonances because of their shape, but the resonances could be damped by stuffing the pipe with fiberglass. After a flurry of popularity, the large drain pipes disappeared.

The early attempts to use ceramic tile may have been too ambitious. Tile is dense and rigid, both useful virtues for a speaker enclosure, but the larger sections come in shapes that are usually awkward, either visually or acoustically, and always, by their excessive weight, physically.

Experience with many kinds of tile enclosures suggests that this material is best used with small enclosures for 5 in. or 6 in. speakers. A 12 in. section of 6 in. (I.D.) drain tile, for example, will have about 0.2 ft.^{3} of volume when the ends are enclosed, just about right for many 5 in. high compliance speakers. Construction is fast and simple. All you have to do is to glue plywood ends on the tile with the upper end cut out for a front mounted speaker (Fig. 11-2). A plywood ring to fit around the speaker is a good idea. It permits true flush mounting and will support a screen or grille cloth over the speaker. You can add a reflector for more direct high frequency distribution, but such reflectors can produce uneven response. You should experiment with one before making a permanent installation.

You can get a slightly larger cubic volume and a firing angle of 45 degrees upward from the floor by using a tile with a bend in it. For a 45 degree angle, you would ask for a "⅛ bend" pipe. The reference angle is 360 degrees, so a 45 degree bend is called a ⅛ bend.

This bent pipe in Fig. 11-3 angles the sound upward and out, or in, as you desire. It can be aimed at a wall surface for reflected sound, or reversed and pointed toward the chief listening area for more direct sound. The internal volume is about 0.25 ft.^{3}, or 7 liters.

PROJECT 14: A CERAMIC TILE SPEAKER SYSTEM

For this we chose a 6 in. speaker with a high compliance and a whizzer cone for the highs, installed in a 2 ft. section of 8 in. (O.D.) flue tile (Fig. 11-4). The tile is longer than desirable, but the internal length can be shortened by gluing a partition in the tile, or by adding a gallon bleach bottle filled with sand. The bleach bottle adds considerable weight to an already heavy enclosure, making the total weight about 70 lbs., so the partition is more desirable if the enclosure is to be moved very often.

The partition serves two functions: it limits the internal length so the enclosure does not resonate like a pipe, and it also brings the internal volume down so that the speaker is not over-damped at resonance. Other speakers can be substituted for the one specified, but you should check the resonance and Q of the speaker first. The Radio Shack 40-1285A speaker had a free air resonance of 53.5 Hz which the tile enclosure, after volume reduction, brought up to about 78 Hz. The bass seems especially solid for a speaker that occupies very little floor space, perhaps because the tile is more rigid than a wooden box. Plans and construction details for this project are shown in Chapter 15.

Fig. 11-3. The first step in using any kind of ceramic tile enclosure is to mark pieces of 3/4 in. material for ends. This "⅛ bend" pipe gives a speaker enclosure with a 45 degree firing angle from the vertical.

STUFFED BOXES

Everybody wants something for nothing, so the idea of getting volumetric expansion by filling a speaker enclosure with damping material is appealing. Of course, you *don't* get the volume expansion for nothing. Some energy is absorbed in the stuffing, making the system less efficient, but this is a trade-off that many people seem willing to make. When stuffing is dictated by the size of the box and the characteristics of the driver, it means that driver damping is inadequate in the box without the stuffing; the apparent loss of efficiency has actually exposed a lack of efficiency of the driver, at least for the conditions under which it is operating.

Numerous published tests indicate that some experimenters have found that stuffing improves transient response and reduces phase shift in ported enclosure systems. Since the Thiele-Small work on ported box design, it has become more apparent that stuffing a ported box can flatten response curves, but only where either the original design was wrong, or where the driver did not have enough inherent damping. Some experimenters have claimed to have improved ported enclosures by stuffing the port, but this practice makes the system act more like a closed box with a higher frequency bass cut-off, and also reduces the ability of the port air to damp the driver at resonance. So unless a ported box is badly mistuned, the port should be clear of any obstructions or of damping material. One rule to remember is that transient response is related to frequency response. The flatter the response curve, the better the transient response will be. Any system whose transient response can be improved by stuffing should be regarded as faulty in the original state. The fault may be too small a magnet, or it may be a box that is too small or carelessly designed.

Fig. 11-4. Project 14. A flue tile speaker enclosure.

Fig. 11-5. Project 15 is compact enough to fit on bookshelves or desk tops.

Stuffing does have its uses, particularly for very small boxes. As a rule of thumb, you should cover the interior walls of medium to large size boxes with 1 in. to 3 in. of damping material, but small boxes should be loosely filled. From a bass resonance that is too high in frequency to internal standing waves, the typical faults of speaker systems are worse in small boxes than in large ones.

PROJECT 15: A COMPACT STUFFED BOX SPEAKER

Here we use a ported box for a small speaker, but we will reverse the usual design procedure so that instead of calculating the right box volume, we will set the box volume first and tune the speaker to the box. Stuffing the box will make it "act larger" than its actual volume. The box shown in Fig. 11-5 was designed to fit on a small book shelf. Its outside dimensions are 8 in. wide × 8 in. deep × 12 in. high. This permits an internal volume of less than 0.25 ft.^{3}, so we must use a 4 in. or 5 in. speaker. Normally, a closed box would be used for such a small system, but in this case, we will go to a stuffed ported box to get more bass range. The port will be left open.

A 5 in. speaker with a cloth roll surround was selected and tested, and showed these characteristics:

$$f_s = 85 \text{ Hz}$$
$$Q = 0.67$$

When the speaker was placed over a 0.1 ft.^{3} test box, there was no significant change in the resonance frequency, which indicated that something was wrong with either the test or the speaker.

Inspection showed that the outer edge of the cloth suspension was not covered with damping sealant and that the open weave fabric was leaking air. This defect can be present in any speaker with a treated cloth surround if the sealant does not cover the entire surface of the cloth.

To correct this leakage, use one of the silicone rubber sealants. Apply as thin a coat as possible with the broad end of a flat toothpick. Do not let the sealant build up on previously treated parts of the suspension, but rake it off while it is fresh. When you have covered the untreated part, put the speaker aside to let the sealant cure for twenty-four hours or more.

This speaker required two treatments to seal the entire edge of the surround. After sealing, the tests showed:

$$f_s = 91 \text{ Hz}$$
$$Q = 0.7$$

Since this speaker was to be used in an enclosure of predetermined size, there was no need to make a V_{as} test. Instead, the cabinet can be tuned by using it as a test box and determining the V_{as}/V_B ratio with the stuffing in the box.

The box was loosely filled with Dacron batting and a 3½ in. length of 2 in. tube installed for a port. When the speaker was installed, tests indicated that the critical frequencies were: f_L, 66; f_B, 93; f_H, 123.

Using the formula:

$$V_{as}/V_B = \frac{(f_H - f_B)(f_H + f_B)(f_B + f_L)(f_B - f_L)}{f_L^2 f_H^2}$$

$$V_{as}/V_B = \frac{(123 + 93)(123 - 93)(93 + 66)(93 - 66)}{(66)^2 (123)^2}$$

$$= 0.4$$

This means that by Keele's hand calculator formulas, the tuned frequency of the box should be:

$$f_B = (0.4)^{0.32} (91)$$
$$= 68 \text{ Hz}$$

And:

$$f_3 = \sqrt{(0.4)(91)}$$
$$= 58 \text{ Hz}$$

This is an amazingly low cut-off for a 5 in. speaker in a box that occupies about a quarter of a cubic foot of space. We must remember that these figures tells us nothing about power levels or distortion.

We naturally would not expect this system to deliver the undistorted acoustical power of a larger speaker in a larger enclosure.

A 5 in. tube tuned the box approximately to the desired frequency. That puts the rear end of the tube a little too close to the back of the enclosure for comfort, but the little speaker sounds good when used at the normal small volume for which it was designed. Enclosure plans are shown in Chapter 15.

SINGLE WOOFER STEREO

Early in the stereo age, a number of audio engineers came up with a novel idea—a stereo speaker system with just one low frequency woofer to handle the bass from both channels. Since low frequency bass in non-directional, they said, a single bass source for both channels will give as good a stereo reproduction as two.

This opens the possibility for some unusual speaker systems with certain advantages and a few disadvantages. The most obvious application is to have two small full range speakers handle all but the lowest bass. The large bass speaker can be hidden somewhere, behind furniture or maybe in a closet, while the compact speakers can be placed in the location that will give the best treble dispersion without regard to bass requirements. These "satellite" speakers can be extremely small if necessary and, if the bass speaker is hidden, listeners will marvel at the great bass response of the little speakers. Their treble dispersion will be better than that of tweeters operating from a wide enclosure.

These theoretical benefits of a separate woofer system can be lost if the crossover frequency is made too high. If the crossover frequency is very far above 100 Hz, the stereo image will be degraded and you will be able to hear a divided soruce effect at frequencies near the crossover point. Ideally a 12 dB/octave network should cross over at 80 Hz, and a 6 dB/octave network at 60 Hz in order to insure that separation at 200 Hz will not be impared.

This brings up the crossover network, one of the problems of this kind of setup. Chokes for conventional crossover networks designed to cut off below 100 Hz, present a problem. Also, there is the matter of feeding both channels of an amplifier through one bass speaker without affecting the amplifier's stability or damaging it. The most practical solution for most systems is to use a separate power amplifier for the bass speaker. If you have or can find an extra single channel power amplifier, you can hook up a simple RC low-pass filter similar to that shown in Fig. 11-6. For a crossover frequency of 80 Hz, the values of the capacitors should be:

$C = 2000/R$

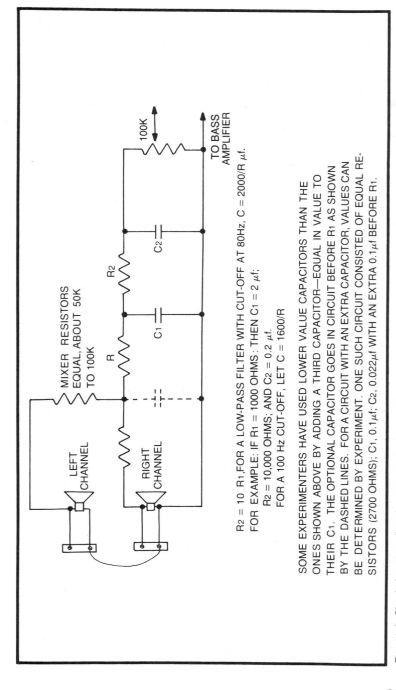

Fig. 11-6. Simple low pass filter for separate bass amplifier.

To prevent interaction between stages, make $R_2 = 10\ R_1$. These formulas should be treated as rough guides. You can experiment with various values of capacitors to tailor the network to your speaker systems. Figure 11-6 shows one slightly different filter with some practical values.

PROJECT 16: LOW COST SUB-WOOFER

If the idea of a single woofer that can be hidden away somewhere appeals to you, here is a possible woofer/enclosure combination, but you may have trouble hiding it. Or you can add a 3-way crossover network with a mid-range driver and a tweeter and have a large, conventional speaker system. Anyway, no set of speaker projects would be complete without at least one big one. To make it more versatile, we are showing it here as a separate woofer so you can use it to extend the bass of your current speaker system if you want to.

Large speaker projects are almost always expensive, but this was a bargain system. We picked up some scrap chipboard in the discard pile from a furniture factory, and found our woofer priced at $15. You can look around for other bargains and use the same design procedure shown here.

Speaker tests showed these values:

$$f_s = 17\ \text{Hz}$$
$$V_{as} = 40\ \text{ft.}^3$$
$$Q = 0.5$$

Some people get 15 in. woofers and then install them in compact boxes, but the V_{as} figure here should prove the futility of that practice. Of course, this speaker has a high compliance foam roll surround that makes it an acoustically big driver, but most modern large woofers are similar in design. The typical V_{as} figures for 15 in. woofers run from 30 to 40 ft.^{3}

Going to the design chart in Fig. 6-8 it looks as though the smallest box size practical for this woofer is a 10 ft.^{3} box, which would put the system resonance at 38 Hz and raise the Q to a little greater than one. There seems no sense in using a 15 in. woofer and letting the system resonance go much above 40 Hz; you can build smaller systems that will go that low. There is one way to shrink the box slightly and still get good performance: use plenty of damping material for volumetric expansion. In this case, we will arbitrarily cut the volume by 30% to make a 7 ft.^{3} box and partially stuff it.

This project was built as planned. The final system resonance was 38.5 Hz, just about what it should have been for an undamped 10 ft.^{3} The frequency of the resonance was too low to use the normal Q test, but an alternate, less accurate, test indicated that the Q is about one. Construction details are shown in Chapter 15.

12

Contemporary Trends in Speaker Systems

In the last few years, the different kinds of drivers and enclosures in use have mushroomed. Instead of simply having variations on a paper cone in a rectangular box, we now have plastic cones, aluminum-coated cones, and even folded diaphragms, cones in omnidirectional towers, on flat baffle dipoles, and in staggered phase-corrected mounting. Along with unusual materials, drivers now assume unfamiliar shapes: radiating surfaces are flat, cylindrical, accordian-like, funnel-shaped, domed, concave, oval, rectangular, and symmetrical.

It is best to evaluate this rich diversity of loudspeakers with an open mind and some skepticism. In some cases, these innovations represent progress, but because most of them cost more than the traditional paper cone speaker in a simple box, each one must prove its worth. A few radically new speaker systems have been sold before the bugs were out, and no system can live up to its promise if it does not work. The ionic "blue glow" tweeter, for example, has been eagerly awaited for decades, but it has been too unreliable for general use. In some speaker systems, the radically new elements appear to have been developed more to permit the manufacturer to advertise something new and different than to advance the art. Something different is fine *if* you are aware that this difference is the main virtue of what you are buying. The current explosion of new speaker models will probably be followed by a shakedown in the coming years. The really practical new speakers will gain a solid place for themselves in the stereo scene, and others will fade away.

NEW KINDS OF DRIVERS

Here is a run-down on some of the unusual drivers that have replaced the traditional paper cone speaker in some systems.

PIEZOELECTRIC TWEETERS

The piezoelectric principle has been previously applied to such mechanical-electrical transducers as the crystal phonograph cartridge, to ultra-sonic sensors, and to underwater use in sonar transducers, but until recently it did not seem practical for loudspeakers. The old crystal phonograph cartridges were notorious for going bad when subjected to excessive heat or humidity. Motorola has developed a lead zirconate-lead titanate polycrystalline material that will operate under such extreme conditions as under water, or at temperatures of 240 degrees Fahrenheit. The tweeters using this material are also electrically rugged, able to take driving voltages up to 35 V RMS; however, their impedance characteristic, which falls instead of rising with frequency, demands a current limiting resistor if the tweeter is to be operated at a high continuous power level at very high frequencies.

In the Motorola tweeters, two lead zirconate-lead titanate discs make a sandwich with a brass separator between them. The ceramic discs are polarized so that one expands and the other contracts when an electrical signal is applied across them. The sandwich, which is glued together with epoxy, dishes in and out in response to the signal. The driver is coupled to a cone and provides its motivating force.

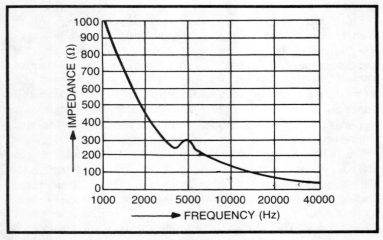

Fig. 12-1. Impedance curve of a piezoelectric tweeter.

Proponents of the piezoelectric tweeter mention that it has such a high impedance to low frequencies, as shown in Fig. 12-1, that it needs no crossover network, permitting the full damping power of the amplifier to be applied to the tweeter. This, and the low dynamic mass of the tweeter, can give excellent transient response, but some users say they have obtained cleaner sound by using a low roll-off filter similar to that shown in Fig. 2-8. The desirability of such a filter depends on the power level of the amplifier and how loud you like your music.

Hugo Schafft, inventor of the Motorola piezoelectric tweeters, has suggested that voice range, or even full range piezoelectric speakers may be feasible if more active and more uniform ceramic materials can be developed.

Unlike some radically new speaker components, piezoelectric tweeters are available to the home speaker builder.

THE HEIL AIR MOTION TRANSFORMER

This driver consists of a folded diaphragm with conductive metal strips on the folds which are located in a magnetic field. As current flows through the strips, they move. Because the metal foil is bonded to the diaphragm, it drives the diaphragm directly, unlike an ordinary dymanic speaker whose voice coil is physically separated from the diaphragm except at its boundary. As the metal strips move, the diaphragm and the pleats in the diaphragm open, and the room air pressure pushes air into the diaphragm. When the pleats close together again, they squeeze air out of the diaphragm. This tweeter has excellent transient response and low distortion.

THE WALSH WAVE TRANSMISSION LINE DRIVER

This driver looks like a funnel, except that the taper is less exaggerated than that of a funnel and is constant, having no break in the angle of the sides. The deep metallic cone can be installed with the "funnel" inverted or upright. It is driven from the small end so that as the wave travels along the cone, the taper, which is angled to match the speed of sound in the cone, provides just the right displacement to make the sound emitted from the larger end of the cone be in phase with that radiated earlier from the small end. The Walsh Wave Transmission Line can give 360 degree radiation from a vertically oriented cone. It is used as a tweeter in several relatively expensive speakers. The Ohm F, which is probably the most expensive single driver speaker you can buy, consists of a large wave-transmission-line radiator which covers the full frequency range.

RTR DIRECT DRIVE ELECTROSTATIC TWEETER

Electrostatic speakers are far from new, but RTR has produced one with some new features. The radiating elements are mounted on a cylinder so that they fire out over a full 360 degree circle. Another feature is the direct drive internal amplifier. RTR says, "All impedance matching components in the internal amp are contained within the feedback loop—cancelling their distortion-producing effects." This internal amplifier is a full range unit, driving a triple woofer system. Integrated amplifier-speaker systems are not in themselves new; for many years some experts have predicted that such systems are the wave of the future. However, so far, they have languished on the sidelines, perhaps because such systems are more expensive but less flexible than independent speakers.

UNCONVENTIONAL PLASTIC DIAPHRAGM SPEAKERS

The first widely available plastic diaphragm speakers were the flat Poly-Planar speakers made by the Magnitran Company. These can be installed in a picture frame, a car door, or any other location where space is at a preimum. Except for the styrofoam-like diaphragm, plastic frame, and flat shape, they are much like a conventional speaker with a magnet and voice coil. The chief problem with these speakers is getting an adequate low bass response when they are used without an enclosure. Measurements on some of these show a high Q, which is all right for their intended use on a flat baffle, or unbaffled.

The Bertagni Geostatic speaker system is a newer system that looks something like the Poly-Planar speakers, but has some important differences. Professor Jose Bertagni, who recently moved from Argentina to California where he has set up a speaker factory, makes a set of drivers that are molded from polystyrene "beads." These diaphragms are asymmetrical, by which design feature Professor Bertagni claims to avoid some of the resonances of symmetrical cones. The thickness of his low frequency driver diaphragms varies; they're thicker in the middle than at their edges. This permits the diaphragm to vibrate in a pulsating motion, with the center moving farther than the edges, which is supposed to produce a fan-like wave. Bertagni says that low frequencies are produced by this pulsating motion, while mid-range frequencies are radiated by the individual beads over the surface of the speaker. Some of the Bertagni speakers have a piezoelectric tweeter for the upper highs.

One of the interesting features of these speakers is their use of mechanical rather than complex electrical crossovers. The voice coil is coupled to the diaphragm with a synthetic material that acts as a frequency filter, depending on the desired range for the driver.

Bertagni says this avoids the phase shift problems of electrical crossover networks.

CONVENTIONAL DRIVERS WITH PLASTIC CONES

Some manufacturers claim that paper is too unpredictable in its physical characteristics to be an ideal speaker cone material. It changes with variations in humidity, the critics say, because it is hygroscopic. They point to experiences where even a change in a single pulp ingredient can alter the frequency response of a diaphragm. But plastic, they say, is a synthetic product which can be more accurately controlled by chemistry. Engineers who favor paper say that the better paper cones are treated with water resistant substances that eliminate the problem of limp cones in damp climates. They admit that the plastic cone is attractive from a manufacturing point of view, preventing problems of variable pulp or other natural substances that can spoil a batch of cone material, but they say that the paper cone at its best can outperform all others. This argument will probably go on until one kind of material wins out in the marketplace, or until some emergency makes one too expensive.

NEW KINDS OF ENCLOSURES

"New" isn't exactly the right word. Many of the so-called new enclosures or baffles are revivals of old ones. They are new in the sense that they were once pretty much abandoned, but have now been resurrected and, in some cases, modified.

THE FLAT BAFFLE

Several manufacturers use no enclosure at all, but mount their drivers on a flat, or almost flat, baffle. Proponents of a flat baffle say that when you put a driver into an enclosure, you produce a resonant peak that colors the sound. All speakers in enclosures, they say, sound like a "speaker in a box." In some models, a group of drivers on a flat baffle cover the full frequency range; in others, just the mid-range is covered.

An example of the latter is the Leslie speaker whose mid-range and tweeter drivers are mounted on a baffle which is shaped like the upper half of an hour glass. The baffle is pivoted so that the user can direct it for the widest stereo coverage from the dipole radiator.

The Bertagni speakers mentioned earlier use the flat baffle arrangement for the full frequency range to obtain what they call "Omnipolar" sound. This is said to give uniform dispersion characteristics for the full frequency range, and to maintain a constant stereo effect over the entire room.

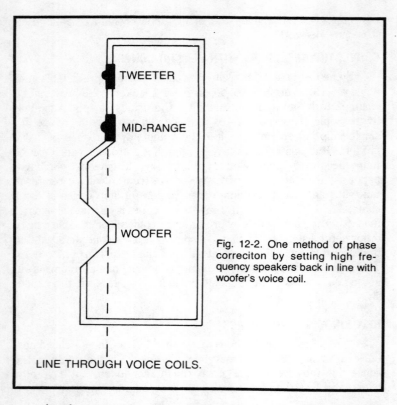

Fig. 12-2. One method of phase correciton by setting high frequency speakers back in line with woofer's voice coil.

Another unusual speaker system that uses an almost flat baffle is the Dahlquist D-10 Phased Array. Except for the woofer, the drivers are mounted on a slightly convex screen. The most important feature of the Dahlquist system is its phase correction, which is a new trend in speakers.

PHASE CORRECTED SYSTEMS

Phase distortion is a controversial subject because authorities do not agree on how much of it can be tolerated. Ever since phase distortion was first discovered in a Hollywood sound studio in the 1930's, it has been a recognized fact that large phase shifts produce unnatural sound. One of the most recent, highly advertised "breakthroughs" in new speaker models has been the staggered physical position of woofers, mid-range speakers, and tweeters. The simplest correction is to move the smaller drivers back in space so their voice coils are vertically aligned with that of the woofer (Fig. 12-2). Some systems use a more complex arrangement to produce a linear phase system by using a time-altered network to correct the phase of the drivers electrically as well as physically. And at least one system, which uses the Time-Align technique, has been used to make the entire phase correction by purely electrical means.

Some of the brand names for phase corrected systems are Dahlquist Phased Array, Technics Linear Phase, Sonic Energy Systems, Sonex, Time-Align, Ultralinear, Bang & Olufsen Uni-Phase, and B & W.

Phase correction appears to have good possibilities, but like any other new development, it should not be the sole criterion in judging or buying a speaker system. After reading the manufacturer's advertisement, one experienced audiophile went to a dealer to audition a new phase-corrected system. When asked how he liked it, he answered that he didn't know because this system had a muddy bass, so he kept noticing that instead of the mid-range and highs.

There is a moral to this: speaker system designers sometimes get hung up on a single sound characteristic to the exclusion and detriment of all others.

TOWERS

While towers have several advantages, they have one disadvantage: if the tower is a pipe-like structure, it can sound very resonant, like a pipe. Some of the tower enclosures are transmission lines, some are reflexes, but others are just elongated closed boxes. Evidently a good many speaker designers think that the tower has enough desirable qualities to off-set the resonance problem.

A tower is an obvious choice for omni-directional sound because it gets the speakers, and especially the mid and high frequency drivers, up where they can fire out all sides into a less obstructed field. A tower can improve dispersion because its narrow frontal area helps to increase the angle of mid-range radiation. The tower occupies less floor space than other kinds of floor speakers of equal volume, which is always a good selling point. Proponents of towers with conventional, front panel speaker mounting say that the tower raises the speakers high enough to prevent floor reflections, and that if the tower is deep enough, it will also reduce wall reflections.

The tower shape may be acoustically controversial, but to some audio fans it offers welcome visual relief from the 3 × 5 boxes.

MISCELLANEOUS UNCONVENTIONAL SPEAKER SYSTEMS

Bose, which has been making Direct/Reflecting speakers for about a decade, has a habit of giving model names that offer a clue to the number of drivers in the system. Their original speaker, the 901 had nine 5 in. speakers in a five-sided enclosure. Eight of the drivers faced the rear on the two angled backs while the remaining speaker faced forward for a direct/reflecting ratio of 1:8. A late model, the

601, has two 8 in. woofers, one in the front and one angled upward, and four tweeters. Three of the tweeters fire at the rear and outside walls while the fourth covers the listening area with direct sound.

Another speaker that radiates sound in several directions is the Epicure Twenty-Plus. The Twenty-Plus has a dual module design, each module containing an 8 in. woofer and a 1 in. tweeter. One pair of drivers, woofer and tweeter, faces foward while the other fires from its tilted speaker board at a point on the ceiling above the heads of the listener. Epicure claims that this arrangement combines first arrival sound from the forward duo with the second arrival sound that has been reflected off the ceiling and walls from the top mounted drivers in a way that simulates concert hall sound.

To end this discussion of unusual speakers, let's take a quick glance at the Marantz Vari-Q systems. These enclosures have a port in them, but Marantz supplies a plug for the port to convert the speakers to closed box operation. Their ads suggest that listeners use the plug for classical music and then pull the plug to get maximum reflex action at about 75 Hz for rock music.

Audio experts might quibble with this idea on a couple of counts. First, they would say, the system should be permanently aligned for best bass performance, either closed box or ported. Anyway—and this point is similar to the first—music is music. A speaker should accurately reproduce the electrical signal, no matter what kind of music is played.

These answers may make theoretical sense, but we might suspect that Marantz has found that rock music fans like a different kind of sound than classical listeners. One wonders if they tuned the box, when ported, for the bass that rock people like instead of trying to get the flattest response. These are only speculations, but without testing one of these speakers, there are some clues that point to this conclusion.

13

Notes on Using Your Speakers

The main rule in setting up a stereo system is to have enough patience to try different speaker locations and various tone control settings. The goal is to choose the set-up that permits you to listen for long periods of time without fatigue. Except for saying that a good speaker will always sound better than a bad one, it is almost impossible to predict how a certain speaker system will sound in a specific environment, but here are some rough guides on how to get the best sound your speakers can produce.

SPEAKER PLACEMENT FOR BEST FREQUENCY RESPONSE AND STEREO IMAGE

The best places in the room for a good frequency response are not always the best for the stereo image. As you probably know, for stereo, the left and right channel speakers should have some distance between them, the amount of distance depending upon the size of the room and how far you sit from the speakers. A plan of your listening room as seen from above should show the left and right channel speakers making an isosceles triangle with the listeners' chairs (Fig. 13-1). The base of that triangle, the side with the speakers at each end, should not be the longest side. If furniture arrangement dictates a wide base, try angling the speakers inward, toward the listening position. If the speaker layout on the enclosure is asymmetric, the side of the cabinet with the tweeters should be on the inside of the triangle for better high frequency coverage of the listening area. This is why a stereo pair of speaker enclosures should

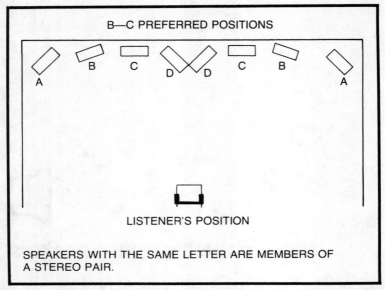

Fig. 13-1. How to arrange a stereo pair of front firing speakers for good stereo image and frequency response.

be mirror images of each other (Fig. 13-2). If you have to put the two channel speakers close together, which would be a rare situation, you may find that facing the speakers outward will give an expanded stereo image, but at the cost of some high-frequency response.

To get good high-frequency distribution, get the tweeters high enough above the floor so that no solid piece of furniture blocks them. Ideally, the tweeters should be on a level with the listeners' ears, although that may not be possible for small floor speakers unless you built a stand to put under the cabinets.

SPEAKER POSITION FOR GOOD BASS REPRODUCTION

Several years ago I built a stereo speaker system that had speakers facing out of each side of two hanging cubes. Almost everyone who heard the system liked the full omni-directional sound, but commented on the weak bass response. Each hanging speaker was radiating sound into a sphere. Because the enclosure was small compared to the wavelengths, the low frequency waves could curl around it. This caused a severe apparent loss of low frequency energy. A speaker on the floor, whether it is in the middle of a room or in the middle of a wall, radiates into a half-sphere. Its bass power is greater than if it were suspended in the air, and the lowest frequencies will be up to 3 dB louder. At the junction of a wall and the floor, which is a typical location for floor speakers, the woofer radiates into a quarter-sphere. The lows are concentrated into a more limited space by floor and wall reflections, boosting the bass another 3 dB. The bass is most powerful when the speaker is moved to a room corner at the floor or ceiling, having up to eight times the level of a suspended speaker, but a speaker in a corner may kick off objectionable room resonances. The cure is to move the speaker away from the wall. When comparing speakers, it is important to remember that bass performance is almost always influenced more by room position and characteristics than by the differences in the speakers themselves.

LEVEL AND TONE CONTROL SETTINGS

Many early hi-fi addicts considered tone control to be superfluous or, when they had to use tone control to compensate for noisy records, a necessary evil. One mark of a hi-fi fan was a flat setting on the tone controls, except for a possible boost in the "presence" range. Sometimes the general effect was similar to drawing a file across a piece of sheet metal, but the owner of such a system called it "hi-fi" because it went higher in frequency than the muffled radios and phonographs of earlier times. Many females objected to the bright highs and, when tests showed that they often had wider range hearing, the practice of running tweeters flat out came under question. Even the most casual listener could tell the difference between the soft natural highs of a live performance and the hard highs of early hi-fi-systems.

Some, but not all of the difference, was caused by distortion in the equipment. Most recordings are made with the microphones close up, which practice preserves the high frequency response and probably adds definition, but it also emphasizes highs. In an au-

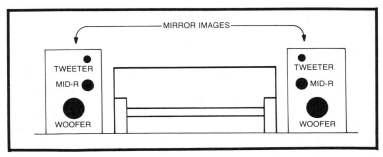

Fig. 13-2. Asymmetrical systems should be made in mirror images of each other. The high frequency speakers should be placed toward the inside of the speaker/listener triangle as shown here.

ditorium, the sound is reflected around the room before it reaches the ear of the listener, a situation which gives excellent dispersion and a falling frequency response. The same thing can happen in a listening room, particularly with an omni-directional speaker, but to a lesser extent because the listener is much closer to the source. If you listen to a speaker system when you are sitting a foot or two away from it, you will notice that the highs seem more dominant than if you were sitting at a greater distance. This suggests that you should adjust the tone controls differently for different situations.

To adjust the tone on your speakers, start with the level controls on your tweeters and mid-range drivers. Set them so the sound appears to be fully integrated without separate sources for treble and bass. For the right overall tonal balance, use the tone controls on your amplifier (Fig. 13-3). Many listeners prefer a slight treble cut, since they consider that setting to be more like live music. As shown above, there is some justification for this. You can often get a smoother bass response by placing the speakers in positions that give reduced bass, and then using bass boost in the amplifier. The best advice is to adjust the controls so that you can listen for long periods without fatigue, and to apologize to no one for their setting.

You may have to make frequent changes in your amplifier tone controls because of the enormous range in program quality, especially broadcast progrmas. Tune any FM receiver across the dial, and, if you're lucky, you will find one or two stations that fulfill the promise of the medium. At other points you'll find distortion, unnatural bass boost, or wiry highs. Until both speakers and program sources are perfect, tone controls are here to stay.

PROTECTING YOUR SPEAKERS

If you have a high powered amplifier, you may worry about your speakers. Assuming a good turntable and other components, you should be able to use high powered amplifiers with no strain, unless you do something foolish such as connecting a low power small speaker to a 200W amplifier and driving it all out. The extra power in the amplifier will give you reserve for momentary peaks without hurting the speakers. In fact, if you run your speakers at rock concert levels, some high powered amplifiers may be safer for tweeters than lower powered models.

These reassurances may be worthless for a particular combination of amplifier and speaker system. The obvious solution is to put a fuse in each speaker line to protect the speakers. Get a fuse holder to match the fuse you need, and mount it on the back of the enclosure. Wire the fuse in series with the speaker system. A fuse can make an audible difference in the sound because fuses have

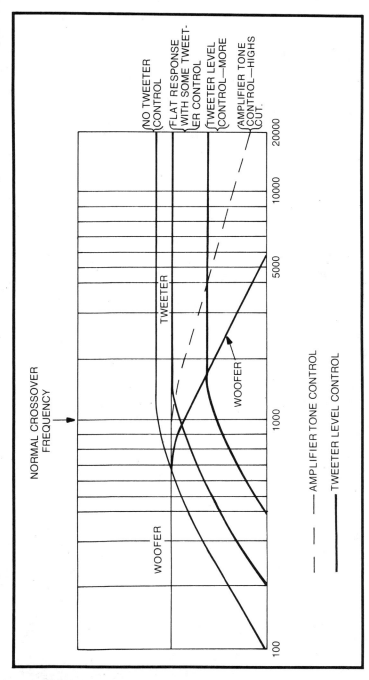

Fig. 13-3. How the action of a tweeter control on the speaker system differs from the treble tone control on the amplifier.

resistance, so they will raise the total Q of your amplifier/speaker system circuit. If the Q is lower than necessary, this can be a good thing; you'll hear more bass. If the Q is marginal on the high side, you may get muddy lows with hangover. The best way to find out is to try fusing your speakers, one at a time, and compare the sound. If the sound is satisfactory with the fuse, leave it in.

To determine the size of the fuse you need, use this formula:

$$I = \sqrt{\frac{P}{Z}}$$

Where:
I = fuse rating in amps
P = power rating of speaker
Z = impedance of speaker

For example, if you want to use an 8Ω speaker system that is rated at 20 W with a high powered amplifier, say a 100 W model, what size fuse would be appropriate?

$$I = \sqrt{\frac{20}{8}} = \sqrt{2.5} = 1.58$$

You can't get a 1.58 A fuse, so you would choose the next smaller size, a 1½ A fuse.

HOW TO HOOK UP EXTENSION SPEAKERS

Some audio fans avoid extension speakers on the grounds that they do not have room for two speakers in each room. If extension speakers are desired for background music, one speaker per room is enough. You can operate the amplifier or receiver in the "mono" mode and feed a mixed signal to each remote location.

Most receivers and amplifiers have a provision for at least one set of extra speakers. In a typical hook-up, you would have your main speakers on the A circuit and an extension speaker on each channel of the B circuit. For example, one extension speaker in a bedroom could be wired to the left channel and another extension speaker in the den to the right channel. This hook-up would cause no problems if the speakers all have an impedance of 8Ω or greater, even though the B speakers are in parallel with the A speakers. The chief precaution to use with an extension speaker circuit is to maintain a minimum load impedance of 4Ω on the amplifier.

If you want to use more than two extra speakers, you will have to add a switch, unless all extension speakers are to work together. Don't forget to check the total impedance. Suppose you want to add a total of four extra speakers. Wire a pair of 16 Ω speakers in parallel

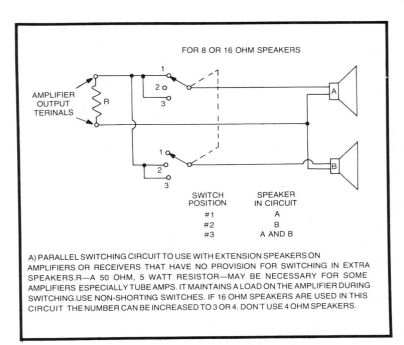

Fig. 13-4. Extension speaker circuit for 8Ω or 16Ω speakers.

to each channel in the B circuit of your receiver. This would give the extension speakers a net impedance of 8Ω. When the A speakers are switched on, the net impedance of the total hook-up will be 4Ω if they are 8Ω speakers (Fig. 13-4).

It is possible to use 4Ω speakers in this situation, but you would need a series circuit for the extra speakers so that their net impedance would be 8Ω. Series circuits should not be used unless the speakers in the circuit are identical.

If you want to use a single pair of 4Ω main speakers, the problem is a little different. You could not use your amplifier switch because a parallel hook-up would bring the net impedance down to 2Ω. Instead, you could use the A circuit only, wiring the extension speakers in a series circuit with your main speakers. A 3-position switch in the circuit of each channel, as shown in Fig. 13-5, would permit you to short out either the main speaker or the extension speaker when not needed.

Switches that are used in output circuits should be of the non-shorting type to prevent their shorting out a solid state amplifier. Some amplifiers, especially tube amplifiers, can be damaged by an open circuit at high power levels, but a 50Ω resistor

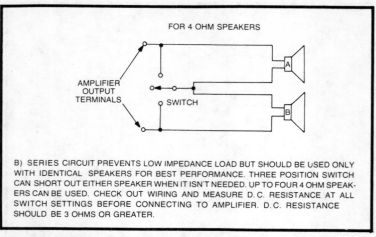

Fig. 13-5. Extension speaker circuit for 4Ω speaker.

across the output, as shown in Fig. 13-4, will prevent the removal of load during switching.

HOW MUCH POWER FOR EXTENSION SPEAKERS?

Rooms that are served by extension speakers normally need less power than a living room. Kitchens, for example, are typically small and hard surfaced. The requirement for other rooms may be decided more by the kind of listening (background music) than by the room's acoustical situation. Outdoor areas require considerable acoustical power, but the speakers designed for outdoor use are usually high in efficiency. As a rule of thumb, you should allow at last 5 W of amplifier power for each extension speaker, and more if you use controls on the speakers to permit the main speakers to operate at full volume.

One item that can affect the power requirement is the length and kind of speaker cable. The gauge number of most wire sold by electronics stores as "speaker cable" runs from #20 gauge down to #24 gauge. Notice that the larger the gauge number, the smaller the wire. The resistance of #24 gauge wire runs about 25Ω per thousand feet, so a relatively short cable of 20 ft. will have a total wire length of 40 ft. and add about 1Ω resistance to the speaker circuit. With a 4Ω speaker system, this much resistance would waste a considerable amount of power and would seriously reduce the amplifier's ability to damp the speaker at resonance. Ordinary #18 gauge lamp cord is the smallest wire that should be used for remote speakers. Figure 13-6 shows two sets of figures for permissible wire length. The first set, the preferred maximum length, will prevent the wire from adding more than 5% to the load. The second set of figures can be used, but with some speakers there may be some loss of quality because the resistance in the wire will increase the total Q of the system.

If you need to use individual controls on the speakers, you can use L-pads or T-pads to balance the sound levels in rooms of different size or acoustical environment. The pads on the speakers will again add some series resistance, and may cause muddiness or boom. For background music, which is the kind of listening for which the L-pads would be used, this may not be a serious fault.

If you decide to wire some extension speakers to your stereo set, don't forget to observe these important precautions:

1. Make sure the net impedance is no lower than 4Ω per channel.
2. Use large enough cable to conserve your amplifier's power and damping ability.
3. Make sure that the speaker circuit can not be accidentally shorted by someone unfamiliar with your switching hook-up.
4. Use non-shorting switches.
5. Use controls only where necessary.
6. Maintain a constant load on tube amplifiers.
7. For series circuits, use identical speakers, if possible.

It is a good idea to measure the dc resistance of your speaker circuits at all switch settings before connecting the speakers to the

	CABLE LENGTH					
	5% LOAD LENGTH			MAXIMUM LENGTH		
WIRE GAUGE	4 OHM	8 OHM	16 OHM	4 OHM	8 OHM	16 OHM
#14	40 FT.	80 FT.	160 FT.	100 FT.	200 FT.	400 FT.
#16	25 FT.	50 FT.	100 FT.	65 FT.	130 FT.	260 FT.
#18*	15 FT.	30 FT.	60 FT.	40 FT.	80 FT.	160 FT.

*#18 WIRE IS ORDINARY LAMP CORD.

CABLE LENGTH IS LENGTH OF DOUBLE CONDUCTOR CORD THAT RUNS FROM AMPLIFIER TO SPEAKER. ACTUAL WIRE LENGTH IS TWICE THE CABLE LENGTH. THE 5% LOAD LENGTH FIGURES WILL LIMIT WIRE RESISTANCE TO ABOUT 5% OF THE TOTAL LOAD, A CONSERVATIVE FIGURE. IF NECESSARY THE LENGTH CAN BE EXTENDED TO THE FIGURES GIVEN IN THE COLUMNS UNDER "MAXIMUM LENGTH." THESE LENGTHS MAY CAUSE A CHANGE IN SOUND QUALITY WITH SOME SPEAKERS. CABLE WITH CONDUCTORS SMALLER THAN #18—SUCH AS #20 TO #24—SHOULD NOT BE USED UNLESS CABLE LENGTH IS VERY SHORT. SMALL CONDUCTOR WIRE IS OFTEN SOLD AS "SPEAKER CABLE."

Fig. 13-6. Cable length for extension speakers of 4Ω, 8Ω, and 16Ω impedance.

amplifier. A dc resistance of less than 3Ω should be checked out for wiring mistakes.

SPEAKERS FOR QUAD SYSTEMS

Many of the first stereo systems consisted of an old hi-fi speaker plus any odd speaker the audio fan could fit into his living room. Later it became apparent that identical speakers gave a more stable stereo image without the shifting sound pattern produced by speakers that had audibly different characteristics. Now, hardly anyone listens to stereo with dissimilar speakers for the two channels. This suggests that the same rule should be applied to setting up a quad speaker system: identical speakers all around.

Of course, if you have two large speakers and want to add two more, things may get a bit crowded. Here is where the speaker builder has an advantage. He can build two extra speaker systems with the same tweeters that were used in the main speakers, but with a smaller woofer in a compact box. If the upper mid-range and highs are produced by identical speakers, the overall sound should blend very well.

The same principles apply to speaker location for quad that apply to stereo, except that you should ideally have a rectangular arrangement with the listening position somewhere in the middle. The only other difference seems to be that with four speakers the odds usually prevail and one of them gets connected out-of-phase. You hear this effect at most electronics stores that sell quad systems. Moral: use Test 13, Chapter 3, for polarity when wiring your system.

14

Rainy Day Projects

Some of these projects can be used with speakers of all grades. Others should be reserved for cheap speakers. Those that involve filters can be undone, but once you tinker with a speaker's cone, the change will be permanent until and unless you ship the speaker out to be reconed. With that caution, here are some ways to tailor the frequency response of your speakers—for the better, we hope.

One typical characteristic of many speakers is a forwardness in tone caused by too much output somewhere in the upper mid-range to lower high frequencies, from 500 to 2000 Hz. This can make a speaker sound loud even at moderate volume. Cheap speakers usually have several bad peaks in their response curves, along with some dips that look just as pronounced on a frequency graph as the peaks, but are not so apparent to the ear. You can improve the sound of many speakers significantly by leveling off the peaks.

HOW TO TAME BAD PEAKS

The first step is to identify the frequency of the peaks by doing Test 9 in Chapter 3. If there are numerous small peaks, ignore all but the most prominent. Minor peaks are not important, and there is a practical limit to how many peaks you can tame in one speaker. You can filter out one or two bad peaks, but if a speaker has more than that, forget it.

When you have identified a peak, record its frequency. If possible, use Test 2 to measure the impedance of the speaker at that frequency. Then use the crossover network design chart in Fig. 2-9 to get the value of a choke and a capacitor that will each have a

253

reactance at that frequency that will be about the same as, or greater than, the speaker's. The exact values may depend on the kind of spare parts you have on hand. If you are unable to run an impedance test, make the reactances twice as great as the rated impedance of the speaker. Wire the choke and capacitor in parallel and put a resistor in parallel with them as shown in Fig. 14-1A. To correct a broad peak, its resistance should be about equal to that of the choke and capacitor at the peak frequency; to correct a sharp peak, it should be much greater. A variable resistance is a good idea; you can try various settings until the sound is right. Try any 50Ω to 500Ω variable resistor, but keep the power low to prevent damage to low wattage resistors. You can measure the resistance that gives the smoothest sound and put in a 5 W to 10 W resistor of that value.

Here is a worked example: a speaker has a bad peak at 1000 Hz. You do not know the exact impedance at 1000 Hz, but the rated impedance is 8Ω. Going to the chart in Fig. 2-9, and following the 1000 Hz line up to 16Ω, which is twice the rated impedance, the diagonal lines that cross there show a 2.5 mH choke and a 10μf capacitor.

The next step is to wind a suitable choke and get a 10 μf capacitor, either a surplus oil-filled capacitor, or a non-polarized electrolytic. Before wiring them into the parallel circuit, wire them into a temporary series circuit with a 16Ω resistor for Test 17 in Chapter 3. This test will tell you their frequency of resonance, which should be close to 1000 Hz.

For better speakers, you should use lower value resistors and effect an overall frequency response correction rather than just a peak correction. Many people like a gentle slump in the response curve in the 1000 Hz region, so a good start for any single cone speaker would be a filter with about 8Ω to 16Ω reactance for each component, and with a 10Ω to 20Ω resistor in parallel.

If you have a reliable frequency response curve for your speaker, you can use the formula in the back of the book to calculate the impedance curve of the filter at various frequencies. Plot the curve on a graph to get an idea of how changing the value of resistance will alter the sharpness of tuning. However, if your choice of capacitor and choke was about right, a variable resistor will tell you the same thing by ear.

If the peak frequency falls between two of the frequencies listed in Fig. 2-9, you will have to interpolate or do some calculations. For example, let's say you have an 8Ω speaker with a peak at 1500 Hz. The chart tells you that the choke value should be between 1.25 mH and 2.5 mH, and the capacitance should be between 5 μf and 10 μf. Since 1500 is midway in frequency between 1000 and 2000 Hz, you may be tempted to get an average value for each, but close inspection will show you that the frequency scale on the chart is not linear. The average values would come out 7.5 μf and about 1.9 mH, which is too high. A filter made with them would have resonance at about 1340 Hz. Instead of estimating the values, it is better to calculate them from:

$$L = \frac{X_L}{2\pi f}$$

$$= \frac{16}{6.28 \times 1500}$$

$$= 0.0016985 \text{ H or } 1.7 \text{ mH}$$

$$C = \frac{1}{2\pi f X_C}$$

$$= \frac{1}{(6.28)(1500)(16)}$$

$$= 0.0000066 \text{ f or } 6.6 \ \mu\text{f}$$

So a tentative filter for this speaker would consist of a 1.7 mH choke, a 6 μf to 8 μf capacitor, and a 16Ω resistor in parallel. This filter should be wired in series with the speaker as shown in Fig. 14-1A. If the filter depresses the mid-range too much, reduce the value of the resistor. If the filter is not effective enough, the resistance should be increased, or the filter frequency changed by varying the values of the components. You can easily make these changes if you have a tapped choke, as shown in Chapter 2, and several capacitors to wire in parallel. Keep the resistance at least equal to or greater in value than the reactance of the other components at the peak frequency. Remember this, a low resistance broadens the filter bandwidth as it reduces the filter's effectiveness.

For many speakers, peaks show up below or around 1000 Hz and again somewhere in the high frequency range. If you can identify the regions of the peak frequencies, you can wire two filters in series, one tuned to each peak.

If you have no equipment for locating the frequencies of a peak in a cheap speaker, you may be able to estimate the approximate range of the peaks by careful listening. For example, if the speaker seems to project instruments forward and sounds louder than it should for the level of sound output, the peak is usually in the upper mid-range, probably around 1000 Hz. If the sound is brassy, try a filter tuned to 2000 Hz. For a speaker that makes record surface noise sound worse than it should, the peak may be higher up the scale, around 3500 Hz. If the sound is sputtery, try 7500 Hz. Even if you can measure the peak frequencies, the final adjustment of the filter should be made by ear. Too much filtering, even if theoretically correct, can make a particular speaker sound dead, but used with common sense, it can sometimes make an inexpensive speaker sound surprisingly good.

R$_x$ FOR TWEETERS

The best cure for an unsatisfactory tweeter is to get a better one but, in the meantime, here are some ways to soften a shrill tweeter or pump up the highs on a dead one. First, analyze your problem. Most tweeter problems can be classified as simply too much or too little. If the tweeter is too prominent and you have no balance control, see if you can move the tweeter to a position that will spread the high response by reflection. This gives better high frequency coverage of the listening area, but some of the upper highs may be lost.

Another, more common cure for a hot tweeter is to put a control in it. An obvious choice is an L-pad that has a resistance in each leg equal to the rated impedance of the tweeter. If you do not want to buy an L-pad for a cheap speaker system, you can use a simple network consisting of two resistors, one in series and one in parallel with the tweeter, as shown in Fig. 14-1B. The resistors in the network should be rated at 5 W to 10 W in speaker systems that will be driven to full volume in average size rooms; 2 W resistors are all right for most extension speakers or other low power systems.

If you have a junk box of odd resistor values, the best way to choose the combination is by trial and error. The chart on Fig. 14-1B shows how to choose values to get a specific change in loudness. For example, if you want to reduce an 8Ω tweeter's loudness by 6 dB, you can use a 4Ω resistor for R$_1$ (0.5 Z), and an 8Ω resistor for R$_2$. If you use a low value resistor for R$_2$, check the load impedance of the speaker at high frequencies.

A dead tweeter poses a tougher problem. One trick is to put another tweeter in parallel with the first, which reduces the impedance and increases power to the tweeters. But you should check the effect on the system impedance at high frequencies to prevent a low impedance load that can damage some solid state amplifers. If the impedance is marginal, say 3Ω at a certain frequency, you can add a 1Ω or 2Ω series resistor in the tweeter circuit; but too much resistance will bring back the original problem.

Another way to balance a system with an inefficient tweeter is to pad the woofer with the same kind of circuit shown in Fig. 14-1B for tweeters, but any resistance in the woofer circuit can raise the total system Q, producing boom and hangover. The best solution is a more efficient tweeter.

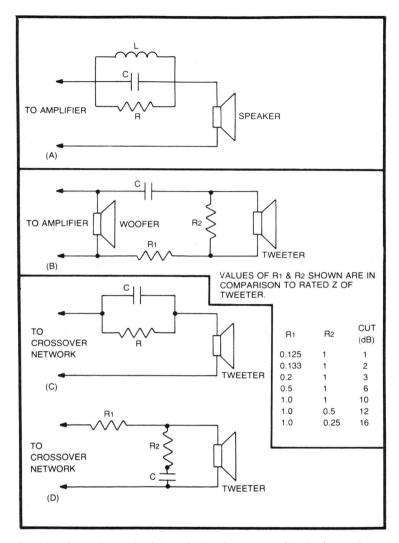

Fig. 14-1. Several ways to change the tonal response of a speaker system.

If the tweeter has too much efficiency in its lower operating range, but falls off rapidly above a certain point, say 10,000 Hz, you can reduce the lower frequencies and boost the highs by putting a parallel RC filter in series with the tweeter as shown in Fig. 14-1C. For an 8Ω tweeter, the resistor should be about 8Ω, and for the boost to start at 10,000 Hz, the capacitor should have a value of 2 μf. Use a larger capacitor to lower the boost frequencies.

If the tweeter has the opposite problem, which is too much top, or a rough top, you can use a series RC filter across the tweeter. However, it is a good idea to insert an extra series resistor, if possible, to prevent a low impedance load for the amplifier at high frequencies. The circuit shown in Fig. 14-1D would have these typical values: R_1, 5Ω; R_2, 1.5Ω to 2Ω; C, 10 μf to 100 μf. Of course, this will almost eliminate the upper highs; but if the tweeter is peaky near the top it is better to lose these frequencies than to reproduce them badly. This same filter can be used with a shrill sounding full range speaker. Just keep the net impedance above the danger level, which is about 3 Ω for many amplifiers.

IMPROVING WOOFERS

A speaker with a light, stiff cone will have a relatively high frequency of resonance and its bass response will roll off below that point. If the speaker is to be used in a simple box, its bass range can only be extended by lowering the resonance frequency. You can do this in two ways: by increasing the compliance or by increasing the mass of the cone. There is a limit to how much you can improve a cheap speaker this way. If you try to drive it at high sound levels, the short voice coil will move out of the region of maximum magnetic field, increasing distortion. For background music, such a speaker will work very well and sound much better than before modification.

HOW TO INCREASE COMPLIANCE

Compliance changes are feasible for speakers with a simple cone, that is, one that has no surround except corrugations in the cone material. Speakers with separate surrounds such as treated cloth, foam roll or butyl roll surrounds should not be altered unless the surround material has deteriorated.

The easiest way to increase compliance is to slit the cone surround with a sharp knife. The slits should be made in pairs, the two slits in a pair being opposite each other. This technique works best on 8 in. to 12 in. speakers, but it can be used on some smaller speakers with success. You can expect to lower the frequemcy of resonance by 15% to 30% by this treatment. The typical number of slits required to do this is usually 16 or 32.

Start by slitting the corrugated edge at one side of the cone; then make another slit opposite the first. Rotate the cone a quarter turn and make another slit; then one opposite it. You will now have 4 slits; one each at 3 o'clock, 6 o'clock, 9 o'clock, and 12 o'clock. Next, make a slit midway between these. Again make a slit between the ones already made, making a total of 16 evenly spaced slits around the edge of the cone, as shown in Fig. 14-2A. If necessary, make another round of slits.

Install the modified speaker in a closed box to add a restoring force to the cone. The slits may cause leakage, but even so the bass response is usually much improved. The slits can be treated with silicone rubber to seal them. Although the sealer will raise the frequency of resonance slightly, performance is more predictable with a sealed cone.

Here is the result of one treatment: a 12 in. speaker had a cone resonance of 45 Hz. If your speaker has a resonance this low, it is questionable whether you should treat it unless you want to make it suitable for a smaller box. After four slits, the resonance dropped to 38.5 Hz. Four more brought it down to 34 Hz, and sixteen slits, to 28 Hz. that was considered low enough for a speaker with a conventional voice coil. After treatment with silicone rubber compound to seal the slits, the final resonance was 32 Hz.

Any treatment that increases compliance will reduce the Q as well as lower the resonance. In this case, the speaker had a Q of about 1.6 before treatment and about 0.77 after the sixteen slits had been added. After treatment by silicone rubber, the Q was increased again to about 0.85. In this case, as with almost all low priced speakers, the magnet was too small to get the Q down into the range of expensive speakers, but the improvement in freedom from boom will usually be noticeable even when the speaker is put into a small closed box.

A more extreme method of increasing compliance is to cut out the outer corrugation entirely and replace it with a more flexible material. Although manufacturers often use special cloth for the purpose, and ordinary cloth can be used, soft chamois works very well. The chamois has the added advantage of damping the high frequency waves that travel out through the cone and so keeping them from reflecting back into the cone and interfering with following waves. The only disadvantage of this method is that it requires more patience.

Start by removing the speaker gasket. Examine the cone to see if there is a felt dust cover in the center. If there is, you can carefully remove this cover by grasping it with tweezers and pulling it away. Directly below the dust cover you will find the pole piece centered within the voice coil. Cut some strips of photographic film and shim the voice coil to lock it in place on the pole piece. If the cone has a dome center, or a flat center made of the same material as the cone, or any kind of center other than felt, then skip this step. If you use care, the spider will hold the voice coil in place anyway.

Begin work on the surround by cutting out four sections at the 3 o'clock, 6 o'clock, 9 o'clock, and 12 o'clock positions as shown in Fig. 14-2B. Each cut-out section should be about 2 in. to 3 in. wide,

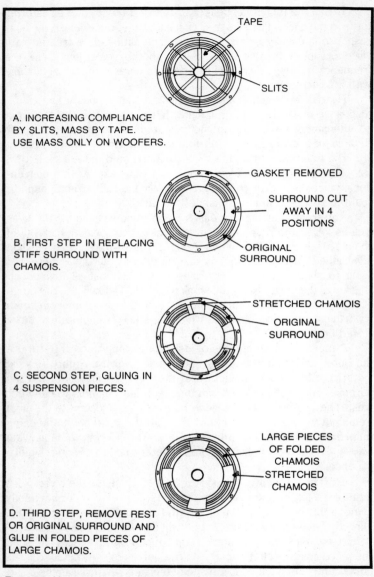

Fig. 14-2. How to increase the mass and compliance of a speaker.

depending on the size of the cone. Cut sections of chamois to suspend the cone; for a 12 in. cone, these sections should be about 2 in. long and wide enough to reach from the edge of the cut cone to the frame. Use rubber cement to glue one edge of each piece of chamois to the lip of the cut cone. When the glue has set, stretch the chamois gently toward the frame and glue it to the frame (Fig. 14-2C). These four equally spaced pieces will suspend the cone and hold it centered. Try the speaker to see if it works all right, but don't drive it hard. If there are no rattles, cut out the remaining cone surrounds and replace them with pieces of chamous. These large pieces should not be stretched, but can be folded so they do not restrict cone movement (Fig. 14-2D). Replace the dust cover if it was removed. There will be some leakage around each of the four small pieces of chamois, but a speaker so treated will work well in a compact closed box.

This method can be used to repair old speakers whose surrounds have deteriorated with age (Fig. 14-3).

HOW TO INCREASE THE MASS OF A CONE

Another way to extend the bass range of a woofer is to increase the mass of the cone. You can experiment with added mass by temporarily sticking a small piece of modeling clay to the cone, but for permanent use, the mass must be glued to the cone. The mass

Fig. 14-3. Radical method of increasing compliance and reducing frequency of bass resonance on old speaker by chamois suspension. Note four sections of chamois that hold cone suspended (photo by D.R. Milne).

should be added symmetrically and in small amounts. This is one alteration that very quickly runs into the law of diminishing returns: too much mass will reduce efficiency more than it lowers resonance.

The easiest way to add mass to a cone is to stick pieces of tape to the cone in a radial pattern, as shown in Fig. 14-4.

A 6 in. woofer with a treated surround had a free air resonance of 95 Hz. Four pieces of ¾ in. tape lowered the resonance to 67 Hz, which was very promising, but when four more pieces were stuck on, doubling the added mass, the resonance dropped only 3 more Hz to 64 Hz. The law of diminishing returns. On a similar speaker, it took 8 pieces of tape to lower the resonance from 120 to 90 Hz. One advantage of using tape is that the radially applied tape helps to stiffen the cone so it can act more like a piston in producing bass.

If you want to stiffen the cone without adding much mass, you can glue lightweight radial "splines" to the cone. Depending on the size of the cone, various items have been used for this pupose: toothpicks, soda straws, and any other material with a high stiffness to mass ratio. To both stiffen the cone and add mass, you can glue sheets of aluminum foil to it. Choose heavyweight foil if you want to lower the cone resonance. Some experimenters have cut cone-shaped pieces of foil to cover the entire cone, but foil can also be used in patches. Make the patches equal in size and attach them in a radial pattern.

Another way to add mass to a woofer cone is to glue a circle of solder to the center of the cone. Any ring glued to the center of the cone should be non-magnetic: don't use iron. Again, test the speaker by sticking on various masses to see how much a given mass will reduce the resonance frequency before you glue it permanently to the cone. And remember, the more you reduce efficiency by adding mass, the more power your woofer will need. A low priced speaker may not be able to absorb enough power to produce the sound level you want, so don't reach too far.

PATCHING DAMAGED SPEAKERS

Punctured cones have caused a great amount of despair, particularly if the speaker was new when the owner rammed the screwdriver blade, or whatever other tool, through it. Unless the puncture is at the center of the cone, it can be patched with almost no change in performance. You can also patch center holes, but do not add too much mass there; it can affect the high frequency response of full range speakers.

For simple tears, glue the parts together again with rubber cement. If you need to reinforce the cone at the torn area, use a nylon mesh material. If a part of the cone is missing, you probably should consider other uses for the magnet. On the other hand, it is possible to patch large holes with a piece of paper reinforced with glued-on nylon; just don't expect a miracle.

WHAT TO DO WITH A DAMAGED WOOFER

If you have a high compliance woofer that has a burned out or stuck voice coil, you can use the cone assembly as a passive radiator. Just knock off the magnet and remove the pole piece. A quick inspection will show the best way to go about this. Most modern woofers have a magnet assembly that is put pogether with epoxy, but some may have rivets that require a chisel. Use a mechanic's ball peen hammer to knock off the magnet. A carpenter's hammer is too hard and may break if used to strike iron parts.

If you damage the cone while converting the speaker to a drone, don't worry about it. Just patch the damage and use it. The added mass may be needed anyway to tune the reflex you will put it in. The chances are you will have to add extra mass by gluing heavy cardboard or washers to the center of the cone.

Low compliance speakers can also be converted to passive radiators by this method, but you must increase the compliance by one of the methods described earlier.

PROJECT 17: A LOW-COST WOOFER-TWEETER SYSTEM

This project grew out of a bargain set of speakers that looked promising, but had a few flaws. A small woofer and tweeter were advertised at $4.00 for the pair. What can you expect for $4.00? In this case a 5½ in. woofer with a treated cloth surround and a 3 oz. Alnico V magnet plus a 2 in. tweeter with a label on the back that lists the frequency range as 3000 to 8000 Hz. A quick frequency test showed that the tweeter goes far above 8000 Hz, which was quite a puzzle (Fig. 14-5).

A visit to a local department store turned up a black grille for a car speaker that measured about 5¼ in. × 10 in., about right to fit the front of an enclosure for the pair of speakers. With that in mind,

Fig. 14-4. Project 17: Applying pieces of cloth tape to add mass to cone.

Fig. 14-5. Project 17: The box is designated to fit department store auto speaker grille. Sound is remarkable for its cost.

and after some tests on the woofer, a 0.2 ft.³ enclosure was designed to fit the grille.

The woofer had a resonance of about 150 Hz in the empty 0.2 ft.³ box. Loosely filling the box with fiberglass brought the resonance down to 130 Hz. The treated cloth surround gave adequate compliance, so any bass extension would have to come from added mass. Eight 1¼ in. strips of ¾ in. tape lowered the system resonance to below 110 Hz. That seemed low enough. Adding too much mass would only have cut efficiency and seriously affected transient response.

When the tweeter was wired in parallel with the woofer through a 4 μf capacitor high pass filter, the tonal balance was terrible. Tweeters are typically more efficient than woofers, but in this case, the lows were hardly audible and the tweeter sounded as if it were overloaded at full volume. Experiments with various filter components showed that the simple resistive network of Fig. 14-1B gave the best performance. The value of the resistors, 10Ω for R_1 and 5Ω for R_2, reveal the degree of imbalance to be about 13 dB. Values of 8Ω and 4Ω, which would be right for 12 dB correction, would have worked just as well, but I had the 10Ω and 5Ω resistors in my junk box. For normal use. 2 W resistors are adequate for this kind of speaker system, making each of the modifications to this project very low in cost, as they should be. If you like more highs, you can tinker further with the network. You can step up the highs by changing the value of the 10Ω resistor to 5Ω. If you want softer highs, leave R_1 at 10Ω and reduce the 5Ω shunt resistor to 3Ω. And so on. The challenge in this kind of project is to get satisfactory sound without spending much money.

Construction plans and details appear in Chapter 15.

15

Project Plans
and Construction Notes

These plans will give you the essential information on how to build a project, but it's a good idea to find the complete discussion of each project in the earlier chapters and read it. That will show you how the project was planned, so that you will have more insight into why the plans turned out this way. If you want to make some changes then, good.

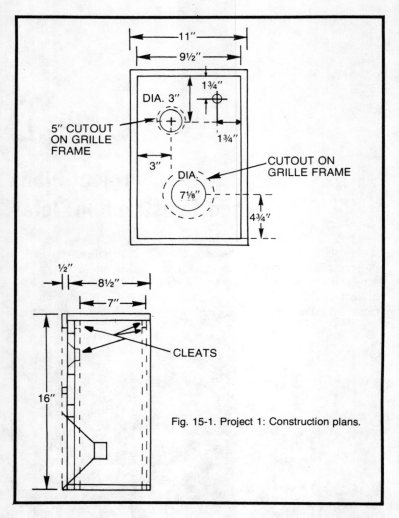

Fig. 15-1. Project 1: Construction plans.

To avoid the common mistakes that many people make in building speaker systems, here is a summary of precautions:

1. Build an enclosure fit for a storm shelter.
2. Seal all leaks with silicone rubber sealant.
3. Test your speakers.
4. Make asymmetrical systems in mirror images for stereo.
5. Get crossover components to match the power rating of your equipment. For high powered amplifiers get capacitors with a voltage rating of at least 50V and L-pads with high power ratings.

6. If you substitute ready-made crossover networks for the homemade ones shown here, try to use similar crossover points.
7. Double check the wiring. Listen to the system at low volume before you seal the box to make sure the highs go to the tweeter and the lows to the woofer.
8. Use the Test 13 for polarity where appropriate.
9. Experiment with the tone controls and the speaker placement.
10. If critical listening suggests a change, *believe your ears*.

The construction notes here may omit certain details, such as painting the speaker board with flat black paint or using glue on all joints except on removable parts, but a re-reading of Chapter 4 will refresh your memory on these universal practices.

PROJECT 1: A SIMPLE 2-WAY SYSTEM

A front view of this project is shown in Fig. 15-2.

Parts List
Enclosure
2	8½ in. × 14½ in. × ¾ in.	Sides
2	8½ in. × 11 in. × ¾ in.	Top and Bottom
2	9½ in. × 14½ in. × ¾ in.	Front and Back
1	10⅞ in. × 15⅞ in. × ½ in.	Grille Frame

Components
1 8 in. woofer Stk. #8E854OCTS
1 1¾ in. tweeter #2TA3
(Speakers from McGee Radio Co.)
1 8Ω L-pad
1 8µf non-polarized electrolytic capacitor
1 0.4 mH choke, about 132 turns of #18 gauge wire, coil form A

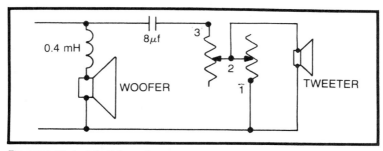

Fig. 15-2. Project 1: Schematic.

Fig. 15-3. Project 1: Wiring details. Here an auxiliary impedance correction filter is connected across the woofer. It consists of a 6.4Ω resistor in series with a 40 μf capacitor. The filter was omitted from the schematic because tests indicated that it was not necessary.

Directions

The construction plans for this project are shown in Fig. 15-1 and the schematic in Fig. 15-2.

The materials for the enclosure can be changed from particle board to plywood, but don't skimp on thickness. Use ¾ in. material as specified. The particle board box shown in Fig. 15-3 was particularly solid and rattle-free.

Cut out the parts and install the cleats. The cleats may not fit together quite as well when you mount them first instead of later, but if you nail the cleats in after the box has been put together, you can weaken the box joints or even break them. If you install the cleats with glue and screws, it will not matter whether they are put on the sides before or after enclosure assembly. Either way you install the cleats, you can fill any small gaps between them with caulking material such as silicone rubber sealant.

Drill two ¼ in. holes at an appropriate distance from each other in the back panel and install a screw terminal strip for the speaker leads. Install the back permanently, using glue.

Drill 11/64 in. shank holes for the screws around the edge of the speaker board at no more than 4 in. intervals. Set the board over the box and drill 5/64 in. pilot holes under each shank hole. Remove the speaker board. Put adhesive backed foam weatherstripping around the board just inside the screw holes where it will meet the cleats. (This enclosure was tested with the baffle screwed down as suggested here. The board was then removed and installed with silicone rubber glue. In both tests the system resonance was the same, which indicates that the weatherstripping made a good seal.)

Install the speakers from the front of the board with a silicone glue gasket under them, and install the L-pad from the rear. Mount a terminal strip between the woofer and the tweeter. Wire the crossover components to this terminal strip, but install the choke about midway between the woofer and the L-pad with silicone rubber glue.

After wiring the crossover network, feed a signal from a receiver or amplifier to the terminals on the back of the enclosure and see if the speakers are working properly. Do the highs come from the tweeter and the lows from the woofer? Is the L-pad working right? When you are sure the hook-up is correct, solder the connections with rosin core solder and install the speaker board.

Unless you built the box with good plywood, cover the exterior with an adhesive-backed plastic veneer. Make a grille frame from ½ in. plywood that will extend out over the entire front of the box with cut-outs for the speakers.

PROJECT 2: A 3-WAY SPEAKER SYSTEM

A front view of the project is shown in Fig. 2-21.

Parts List
Enclosure

2	15 in. × 33¾ in. × ¾ in.	Sides
2	15 in. × 21¼ in. × ¾ in.	Top and Bottom
2	19¾ in. × 33¾ in × ¾ in.	Front and Back
1	7 in. × 12 in. × ⅜ in.	Tweeter Board
1	4 in. × 6 in. × ⅜ or ½ in.	Control Panel
1	19⅝ in. × 33⅝ in. × ⅜ in.	Grill Frame
2	3¾ in. × 4¼ in. × ½ in.	For duct,
2	4¾ in. × 4¼ in. × ½ in.	3¾ in. square
1	30 in. length of 2 by 4	Brace on back

Approximately 24 feet of 1 in. pine cleats

Components
Woofer Sk. #EM-40 WOOF
Mid-R. Sk. #KO4OMRF

Tweeter # K010DT (Speakers available from McGee Radio Co.)
Choke, 2.5 mH, about 350 turns of #18 gauge wire, coil form B
Choke, 0.25 mH, about 115 turns of #18 gauge wire, coil form A
Capacitor, 40 µf, non-polarity
Capacitor, 4 µf, non-polarity
2 L-pads, 8Ω

DIRECTIONS

The construction plans for this project are shown in Fig. 13-4 and the schematic in Fig. 13-5.

Like most of the other enclosures in this book, this was put together from ¾ in. high-grade particle board (Fig. 15-6). Any enclosure this large should have internal corner glue blocks to brace the corners. You may notice that although the inside dimensions of this box do not quite match the "golden ratio," the external measurements do.

The box construction is conventional, but you must use some care in laying out the speaker board. Start out by building the box shell without a front or back. Then set the speaker board, with its

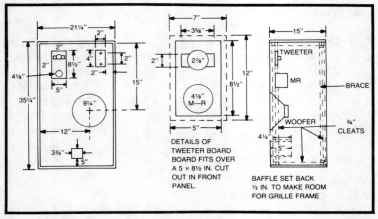

Fig. 15-4. Project 2: Construction plans.

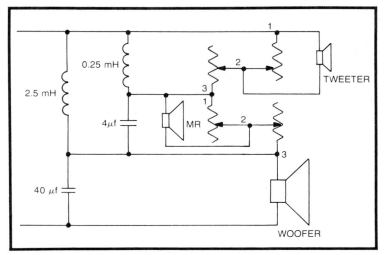

Fig. 15-5. Project 2: Schematic.

speaker holes already cut, into the box. Reach through the back of the box to mark an outline of the cleats on the back of the speaker board, particularly in the two upper corners near the mid-range and tweeter section and the control board. Cut a tweeter and mid-range board and the control panel. Before making holes in them, tack them temporarily in place behind the speaker board. Make sure they do not overlap the marks outlining the cleat positions. Mark the outline of the speaker board cut-outs on the front on the tweeter board and the control panel. Remove the tweeter board and the control panel and cut them for the speakers and the L-pads.

To make the tweeter cut-out, first draw a circle that is 2⅞ in. in diameter. Then draw a 2 in. × 3⅝ in. rectangle over the circle so that the long dimension of the rectangle will be horizontal when the tweeter board is installed in the enclosure. Install the tweeter board and the control panel behind the speaker board with glue, using the previously made nail holes as guides, and install the tuning duct. Drill shank holes for mounting screws and install the foam weatherstripping around the board just inside the screw holes. Install a terminal on the back, and mount the back with glue. Cover the box interior with 2 in. of fiberglass damping material.

Install the speakers with silicone rubber sealant behind each one. Then turn the board face down on a smooth surface to mount the crossover components and wire them. Mount the two chokes with silicone rubber glue.

After wiring the crossover network, set the baffle upright against a bench or table, placing something heavy against the front of

Fig. 15-6. Project 2: Installation of speakers and components on speaker board. Notice use of weatherstripping tape to seal enclosure on both Projects 1 and 2.

the base to hold it. Connect the output of a receiver or amplifier to the system. Listen carefully at low volume while you adjust the controls to make sure that the highs come from the tweeter and lows from the woofer. Then solder the connections, including the lamp cord from the back of the enclosure. Install the front with screws, and cover the box with a plastic veneer. Make a grille frame to fit inside the box sides with the grille cloth stretched over it.

PROJECT 3: A 3-WAY CLOSED BOX SPEAKER SYSTEM

A front view of this project is shown in Fig. 6-7.

Parts List
Enclosure

2	12 in. × 24 in. × ¾ in.	Sides
2	12 in. × 15 in. × ¾ in.	Top and Bottom
2	13½ in. × 22½ in. × ¾ in.	Front and Back
1	14⅞ in. × 22⅜ in. × ½ in.	Grille Frame
1	9 in. length of 3 in. I.D. Tube	
	or	
1	9 in. tube of 2⅝ in square cross section.	
1	3½ in. × 6 in. × ½ in. Control Panel	

About 16 feet of 1 in. pine cleats ¾ in. × ¾ in.

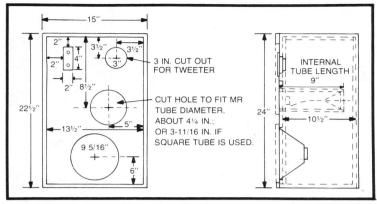

Fig. 15-7. Project 3: Construction plans.

Components
Woofer, Radio Shack 40-1331
Mid-range, Radio Shack 40-1197
Tweeter, #2TA3, McGee Radio Co.
3-Way Crossover Network, Radio Shack 40-1294
2 L-Pads 8Ω, Radio Shack 40-980, or equivalent

Directions

The construction plans for this project are shown in Fig. 15-7 and the schematic in Fig. 15-8.

Make the box of ¾ in. plywood or chipboard. Install cleats all around the front and back and use glue blocks in the corners. Put the back on with glue, but make the front panel removable to service components (Fig. 15-9). Line all interior surfaces except the speaker board with fiberglass batting.

Cut the holes in the speaker board and glue in the tube for the mid-range speaker. Try to cut out the hole for the mid-range tube

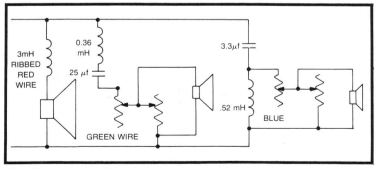

Fig. 15-8. Project 3: Schematic.

Fig. 15-9. Project 3: Internal view. Fiberglass was left off part of interior for this shot so the details of cabinet construction would show.

without drilling a large hole to start the saw, and save this plug. Get a 4 oz. plastic funnel and run a 24 in. length of lamp cord through it, letting about 13¾ in. extend out of the top of the funnel. Set the funnel in a can and fill the funnel with plaster of paris mix. Adjust the lamp cord so that it emerges from the wet plaster of paris at the center of the funnel. Prepare a back for the mid-range enclosure by drilling a small hole for the lamp cord through the center of the plywood and glue the plywood to the funnel with silicone rubber sealant (Fig. 15-10). Fill the hole around the lamp cord with sealant. Then glue the back to the tube with silicone rubber sealant, and glue the tube into the baffle.

Loosely fill the tube with fiberglass or polyester batting. Cut off the lower solder lugs on the mid-range speaker so that it will fit into the tube. Set the speaker board on the cabinet to install it and the other speakers. Install the control panel on the back of the speaker board over the rectangular cut-out and mount the controls on it. Screw the crossover network to the braces with a couple of brass screws.

Note the phasing of the drivers by using the ribbed wire from the crossover network and the red dots on the driver terminals.

Check the polarity of the lamp cord leads from the mid-range speaker with a battery or damper tester (Test 19, Chapter 3). Then connect it to the network. Add about a ½ lb. of Dacron batting to the interior of the box, arranging it so that it will not occupy the space where the mid-range tube belongs. If you use a woofer with a heavier magnet, omit the Dacron filling. Connect the remaining wires to the speakers and controls and test them for proper operaton before installing the speaker board.

Staple the grille cloth to the grille frame and attach it with three or four small pieces of Velcro. The cabinet shown in the pictures was designed to have the grille frame cover the entire front of the box, including the raw edges of the plywood. If you would prefer a different style, you can use hardwood trim to cover the plywood end grain and make the grille frame sit inside the trim.

The final Q of this system should be a little greater than 1, maybe even 1.25 or so, and the system resonance about 60 Hz. This

Fig. 15-10. Project 3: Install the back on the mid-range enclosure with silicone rubber glue. Funnel helps to break up reflections and prevents pipe resonance by providing a tapered tube length.

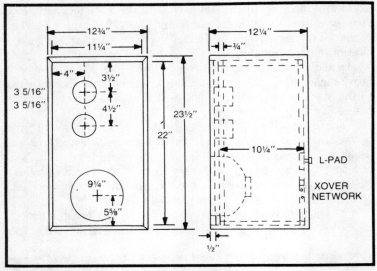

Fig. 15-11. Project 4: Construction plans.

produces a slight prominence for the lower bass, but male speech will be natural and without false boom.

PROJECT 4: A 2-WAY CLOSED BOX SPEAKER SYSTEM

Parts List

Enclosure

2	12¼ in. × 23½ in. × ¾ in.	Sides
2	12¼ in. × 12¾ in. × ¾ in.	Top and Bottom

Above 4 pieces with ends cut at 45 degrees

2	11¼ in. × 22 in. × ¾ in.	Front and Back
16 ft. of ¾ in. × ¾ in.		Pine Cleats
1	11⅛ in. × 21⅞ in. × ⅜ in.	Grille Frame

6 ft. of ¾ in. wood veneer tape

Components

1   Woofer, Cletron, 20 oz. magnet 10 in.
2   Tweeters, C-3½-8
      (Speakers from McGee Radio Co.

1   0.5 mH choke, about 140−150 turns, #18 gauge wire, form A
1   8Ω L-pad
1   12µf non-polarized Capacitor.

Directions

The construction plans for this project are shown in Fig. 15-11 and the schematic in Fig. 15-12.

The box for this project has 45 degree cuts on the sides; with hardwood plywood there will be no end grain showing. If you want to use cheaper material and butt joints, you should shorten the sides to 22 in. to keep the same internal volume.

Cut out the parts and attach the internal cleats and glue blocks. For assembly, coat the beveled edges of the sides, both top and bottom, with just enough glue to do the job and not so much that it will run out and mark the exterior of the box. To hold the parts in close contact while the glue sets, you can assemble the shell with screws through the interior corner blocks, or else with straps around the outside of the box.

Install the speaker board either by gluing and screwing it, or by nailing it to the cleats behind it. When the glue has dried, caulk the joints. Glue strips of ¾ in. wood veneer tape to the front edges of the box. Stain and finish the enclosure.

Next, install the tweeters with silicone rubber glue. Then wire the two tweeters in parallel and in phase with each other. Leave leads from the tweeters long enough to reach the lower part of the back.

Make a 3 in. × 4 in. cut-out in the back behind the woofer position. Install the crossover components with a terminal strip and L-pad on a 6 in. × 8 in. piece of ¼ in. Masonite. Glue and screw the Masonite to the inside of the back so that the L-pad control will fit into the cut-out on the back panel, and wire the crossover network. Line the walls of the box with about 3 in. of fiberglass. Install the back with glue and nails or screws. Caulk the joints between the back and the other parts by reaching through the woofer cut-out. Connect the

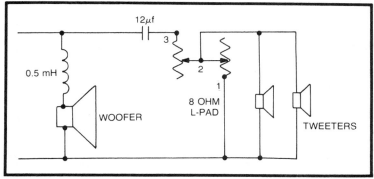

Fig. 15-12. Project 4: Schematic.

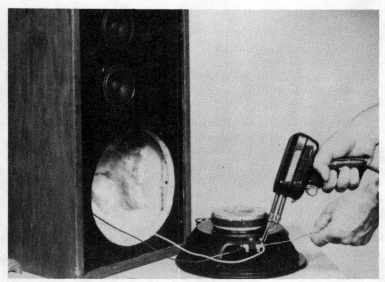

Fig. 15-13. Project 4: The final step is wiring and installing the woofer. Note color coded wire to make phasing easy.

wires from the tweeters to the appropriate lugs on the L-pad. Bring the wires from the choke and the other side of the speaker line out of the woofer cut-out. Cover the back of the enclosure with about 3 in. of fiberglass by rolling a batt of the fiberglass and feeding it through the woofer cut-out. Make slits in the fiberglass to let it fit over the wires from the crossover network, then glue or staple the fiberglass to the back.

Connect the wires from the crossover network to the woofer and test the system. If that is O.K., install the woofer with screws and a silicone rubber gasket (Fig. 15-13). Make a grille frame from ¼ in. plywood and stretch a grille cloth over it. Install it with brads or with patches of Velcro.

PROJECT 5: A COMPACT CLOSED BOX SPEAKER SYSTEM

Figure 6-10 shows this project when completed.

Parts List
Enclosure

2   6¾ in. × 13¾ in. × ¾ in.   Sides
2   6¾ in. × 9⅜ in. × ¾ in.   Top and Bottom
Above pieces cut at 45 degrees on ends.
2   7⅞ in. × 12¼ in. × ¾ in.   Front and Back
About 80 in. of ¾ in. × ¾ in. pine cleats
About 4 ft. of ¾ in. veneer.

Components
Woofer, Norelco 7065/W8
Tweeter, Norelco AD1060/T8
Choke, 0.8 mH, about 210 turns of #22 gauge wire on coil form C.
Choke, 0.5 mH about 170 turns of #22 gauge wire on coil form C.
Capacitor, 20 µf, non-polarized
Capacitor, 8 µf, non-polarized.
The two chokes can be wound from 1 roll of Radio Shack #22 gauge magnet wire, Stock No. 278-003.

Directions

The construction plans for this project are shown in Fig. 15-14 and the schematic in Fig. 15-15.

Cut out the parts according to the plan, making 45 degree cuts at the ends of all the side panels. Install the cleats on the panels, measuring carefully to avoid having large gaps or, even worse,

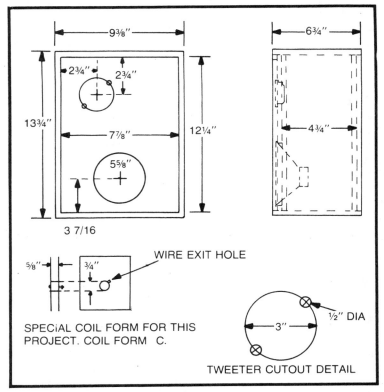

Fig. 15-14. Project 5: Construction plans.

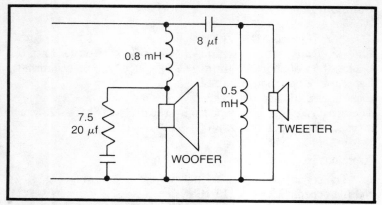

Fig. 15-15. Project 5: Schematic.

having them interfere with the box assembly. Coat the bevel cut ends with glue. Make sure you get enough glue to cover the matching surfaces, but do not use so much glue that it will run out over the exterior. Now set the parts together on edge, circle them with canvas straps or ropes, and pull tight the strap buckles or rope knots. This will hold the parts in close alignment until the glue sets.

While the glue is setting, prepare the front speaker board and the back. If you are using the Norelco AD 160/T8, you will notice that the tweeter mounting hole is unusual. First use a compass to draw a circular outline for the tweeter hole that is 3 in. in diameter. Then draw a diameter across the circle at a 45 degree angle to the sides of the speaker board. At both points where this diameter intersects the circle, drill a ½ in. hole. Cut out the circle and try to install the tweeter from the front of the board. You will probably have to use a wood file or a knife to remove some more material from the "ears" of the tweeter hole in order to accommodate the tweeter lugs. Complete the speaker board by cutting out the woofer hole. Solder short wire leads to each tweeter lug before you install the tweeter.

Install the back with glue, and line the enclosure interior with fiberglass damping material. Set the speaker board in the box to install the speakers. Use silicone glue under both drivers. Although you can use ½ in. panhead sheet metal screws to install the tweeter, the screw heads can damage the butyl rubber suspension on the woofer. A better method is to run a bead of silicone rubber around the mounting hole and to set the woofer on that. Place a small board over its frame and put a small weight on it, just something that weighs a few pounds and will press the speaker frame down into the silicone rubber to make a tight seal.

When the silicone rubber has set, preferably overnight, remove the speaker board and install a 5-terminal strip on the back (Fig. 15-16). Mount the crossover network, and test the speaker at low volume. If everything works right, install the speaker board either with screws and weatherstripping, or with silicone rubber sealant.

PROJECT 6: ELECTRO-VOICE MC8A PORTED SYSTEM

Figure 7-14 shows this project complete and in use.

Parts List
Enclosure

2	13 in. × 28¾ in. × ¾ in.	Sides
2	13 in. × 19¼ in. × ¾ in.	Top and Bottom
2	17¾ in. × 28¾ in. × ¾ in.	Front and Back
1	16 in. length of 1 by 4 pine bracing	For Speaker Board
1	25 in. length of 1 by 4 pine bracing	For Back
1	17⅝in.× 28⅝in. × ¼ in. Masonite or plywood	Grille Frame
2	6 in. × 2¾ in. × ½ in. plywood	For Duct
2	5 in. × 2¾ in. × ½ in. plywood	For Duct

About 20 feet of ¾ in. × ¾ in. pine for cleats.

Fig. 15-16. Project 5: Wiring details. Lower choke is in series with woofer; upper choke shunts the lows across the tweeter.

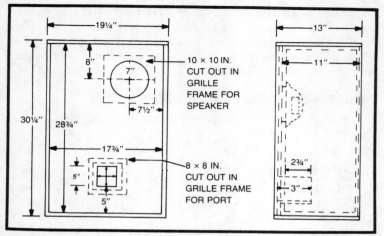

Fig. 15-17. Project 6: Construction plans.

Components

1 Full range high fidelity speaker, Electro-Voice MC 8A
NOTE: This model is not interchangeable with the older MC8. The MC 8A has an improved magnet so enclosure design is different for the two speakers.

Directions

The construction plans for this project are shown in Fig. 15-17.

The wide die-cast frame of the MC8A is clearly superior in damping and stability to stamped frames, but its design does require rear mounting. This suggests a removable back for speaker access. Use weather-stripping tape between the back and rear cleats to seal the box. Apply the tape to the interior of the back just inside the screw holes, where it will compress between the back and the cleats without interfering with the screws.

Round off the front edges of the speaker cut-out with sandpaper or a rounded rasp. Center the speaker on the hole and mark the locations for the flathead 1½ in. × 3/16 in. mounting bolts (Fig. 15-18). You could use T-nuts on the front of the board, but that might require cutting off bolts or using washers to keep them from protruding in front.

Notice that the speaker board is set back ½ in. to make space for the grille frame which is set inside the front edges of the enclosure sides. If you want to use a grille that overlaps the sides, you can eliminate this ½ in. space.

In all other ways, the construction is typical of good quality enclosures; all matching surfaces except the back are glued.

Electro-Voice recommends covering the top, one side, and the back with damping material. Theoretically, that arrangement should suppress mid-range reflections; but after listening tests, we covered all the wall with 1 in. material and the back with 2 in. material.

Tuning

When a 5 in. × 5 in. port with a 4⅞ in. duct was installed in an undamped box, the tuned frequency was about 48.5 Hz, which was very close to expectations. However, this tuning was done by the small speaker method, Test 7, Method IV, with no speaker or duct in the box. After enough bricks had been added to the enclosure to approximate the cubic volume of the speaker and the duct, the frequency was about 50.5 Hz. So the box should be tuned 3.5 Hz higher. Using the tuning correction formula:

$$L_v = -3.5 \ (9.75/50.5)$$
$$= -0.676 \text{ in. (about } 11/16 \text{ in.)}$$

Fig. 15-18. Project 6: Installing Electro-Voice speaker. Tighten nuts snugly to compress gasket, but do not use heavy force, or you might bend frame.

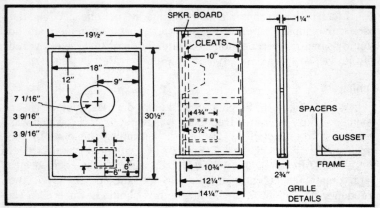

Fig. 15-19. Project 7: Construction plans.

After 11/16 in. was cut from the cut length, the tuning was at 53 Hz. This was judged accurate enough, but when the damping material was added to the box, the frequency was again lowered to 50 Hz. So back to the formula:

$$L_v = 4 \text{ Hz } (8.4/50)$$
$$= -0.672, \text{ or about another } 11/16 \text{ in.}$$

This makes the final duct length about 3½ in., or the thickness of the board plus a 2¾ in. extension behind the board. This should tune the box to 54 Hz if the same amount of damping material is used as in the original: 1 in. to 2 in. on the walls, and at least 2 in. on the back. If you use less damping material, you should add ½ in. to 1 in. to the duct length.

As in the following project, this box volume may seem large for an 8 in. speaker, but the bass response is worth it for anyone who can spare the space.

PROJECT 7: JBL PORTED SYSTEM

Figure 7-15 shows this project when completed.

Parts List
Enclosure

2	29 in. × 12¼ in. × ¾ in.	Sides
2	19½ in. × 14¼ in. × ¾ in.	Top and Bottom
2	29 in. × 18 in. × ¾ in.	Speaker Board and Back
4	28¾ in. × 1 in. × ¾ in. pine	Grille Frame
4	17¾ in. × 1 in. × ¾ in.	Grille Frame
6	1 in. × 1¼ in. × ¾ in. pine	Spacers
8	1 in. × 1 in. × ¾ in. pine	Gussets
About 24 feet of 1 by 2 pine		Cleats and Braces

Components
1 JBL/LE8T Full Range High Fidelity Loudspeaker

Directions

The construction plans are shown in Figs. 15-19 and 15-20.

The enclosure shown in Fig. 15-21 was made of ¾ in. particle board panels with reinforced butt joints. If you plan to use hardwood plywood, you can get a better appearance by cutting the side and top panels at a 45 degree angle at the top for beveled joints and letting the sides extend far enough down to cover the edges of the bottom piece. To do this, you will have to juggle dimensions slightly to get the same internal volume.

The JBL specifications call for 1 in. × 1 in. pine cleats and braces throughout. Lumberyards stock pine in ¾ in. and 1½ in. thicknesses which, in the trade, are called 1 in. and 2 in. boards. So instead of getting 2 by 2's and cutting off the excess, which would have been wasteful, or using them as they were, which would have added more weight, I used 1 by 2's which actually measure ¾ in. × 1½ in. Install the side cleats to take screws from the front and back with their wide surface against the enclosure top, bottom, and sides. Around the middle of the box, where they serve as braces, install them on edge for maximum stiffness. This line of braces around the middle does not match the theory of running all braces lengthwise to the panel; however, as used here, it does have the advantage of tying the back to the two sides at the middle, which may be more effective in strengthening the cabinet as a whole than the best individual panel treatment would be.

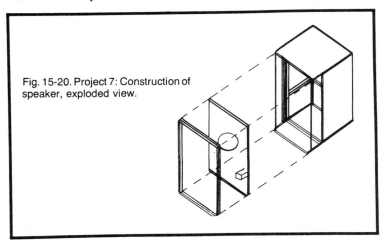

Fig. 15-20. Project 7: Construction of speaker, exploded view.

Fig. 15-21. Project 7: Internal construction. Line all interior surfaces, except the speaker board, with acoustical fiberglass.

Cut out the parts according to the dimensions in the drawings and assemble the sides, top, and bottom. Install all the cleats, including the middle back brace. Then mount the back with glue to seal and strengthen the box. Line the interior of the box with a 1 in. thickness of acoustical fiberglass.

Install the duct on the speaker board. If you plan to fine tune the box to your speaker, you can make the port area slightly larger than it is shown in the plans, and add a lining to reduce its cross-sectional area by gluing in new interior walls of ¼ in. plywood.

Install T-nuts on the rear of the speaker board to match the hole position in the speaker frame. Use glue to mount the board on the box, for the speaker comes with a plastic gasket to permit front mounting. Apply a veneer of your choice to the particle board.

The grille frame is more elaborate than most, constructed of ¾ in. × 1 in. pine material with spacers to give the effect of a deep grille. If you have access to a table saw, you can cut grille frame parts from 1 in. (actually ¾ in.) pine material of any width, but you will have less waste from boards 1 in. × 8 in. or wider.

Use glue to assemble the grille frame. The glued parts should be put together under pressure applied either by screws or by small nails with C clamps. Select an open weave grille cloth that will wrap around the frame edges. With a little stretching, about 24 in. × 36 in. will do.

PROJECT 8: A COMPACT PORTED BOX SPEAKER

This project is shown complete except for its grill in Fig. 7-16.

Parts List
Enclosure

2	7½ in. × 15½ in. × ½ in.	Sides
2	7½ in. × 10½ in. × ½ in.	Top and Bottom
2	9½ in. × 15½ in. × ½ in.	Front and Back
1	2½ in. length of 2 in. I.D. cardboard tubing or frozen orange juice can.	Duct
1	9⅜ in. × 15⅜ in. × ⅜ in. Cut Out to fit speaker and port.	Grille Frame

Components
5 in. × 7 in. Full Range Speaker, Olson Stock No. SP-245

Directions

The construction plans for this are shown in Fig. 15-23.

Small enclosures such as this can be made of ½ in. particle board and assembled without cleats or glue blocks (Fig. 15-24) if you

Fig. 15-22. Project 7: Detail of grille frame corner.

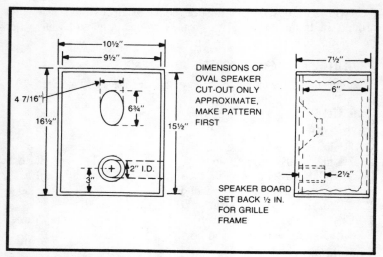

Fig. 15-23. Project 8: Construction plans.

Fig. 15-24. Project 8: Enclosure minus duct and damping material. Speaker is front mounted.

follow the procedure below. Notice that all joints should be sealed with caulking material, and that the parts must fit together closely for good strength. You can follow this general plan for building any small enclosure with a single speaker:

1. Cut out parts and test their fit.
2. Glue and nail a side to the front panel.
3. Glue and nail the bottom panel to the side and front.
4. Glue and nail on the second side.
5. Glue and nail on the top.
6. Caulk all internal joints.
7. Install the duct, unless you plan to tune the box yourself.
8. Glue and nail in the back with the speaker wire installed.
9. Reach through the speaker hole to caulk around the back.
10. Install veneer.
11. Install the speaker.
12. Install a grille frame with a grille cloth.

PROJECT 9: A UTILITY BASS REFLEX SPEAKER

This project is shown complete except for its grill in Fig. 7-19.

Parts List
Enclosure

2	24 in. length of ¾ in. material, tapered from 11⅞ in. to 9⅛ in.	Sides
1	15 in. × about 24¼ in. × ¾ in.	Speaker Board
1	15 in. × 24 in. × ¾ in.	Back
1	9⅛ in. × 16½ in. × ¾ in.	Top
1	11⅞ in. × 16½ in. × ¾ in.	Bottom
1	14⅞ in. × 24 in. × ¼ in.	Grille Frame
2	5 in. × 7 in. × ½ in.	Duct
2	5 in. × 3 in. × ½ in.	Duct
18 ft. of ¾ in. × ¾ in. pine		Cleats & Glue Blocks

Components
Norelco AD 1065/M8 10 in. speaker

Directions

The construction plans for this project are shown in Fig. 15-25.

You can build this box (Fig. 15-26) exactly as planned, or you can avoid the sloping panel by having a rectangular shape all around. To make this as planned, you should saw the top and bottom cleats so that their front edges will be angled at about 7 degrees from a vertical cut. If you use a power saw, set it at 7 degrees and rip both

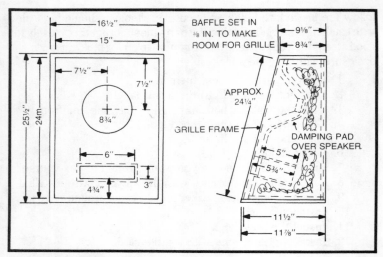

Fig. 15-25. Project 9: Construction plans.

cleats at once, but do not forget to reset the saw at 90 degrees before removing the second cleat from the material.

The construction is otherwise straightforward. Round off the edge of the speaker cut-out to minimize the diffraction from the

Fig. 15-26. Proejct 9: Interior view. In final tuning the duct area was reduced slightly and the speaker covered with a damping pad.

sharp edges. Cover the inside of the enclosure with a 2 in. or 3 in. thick layer of fiberglass. Install the speaker, connect a 24 in. piece of lamp cord to the terminals, and solder them. Select a fiberglass pad at least 1 in. thick and large enough to cover the speaker, and make a small hole in the lower center for the lamp cord. Thread the lamp cord out through the hole and staple the damping pad to the speaker board, stretching it tightly over the back of the speaker. Bring the lamp cord out through the rear panel by any of the air tight methods discussed in Chapter 4. Screw down the back with a screw every 4 in. to 5 in. around its periphery. Test the speaker for proper damping with the damper tester of Test 19.

You can experiment with a cover over the port to compare the performance of the closed box with that of the reflex.

If this speaker were to be used in a stationary box, it would be desirable to make the box larger, perhaps by 50%, to extend the bass range. Make the internal dimensions about 17 in. × 27½ in. × 11 in. or use any other appropriate ratio that will give a volume of 3 cu. ft. Tune the box to the speaker's free air resonance; a 5 in. × 5 in. duct about 3½ in. to 4 in. long would be about right.

PROJECT 10: AN OMNI-DIRECTIONAL SPEAKER SYSTEM

Figure 9-2 shows this project complete and in use, serving double duty as a lamp table.

Parts List

Woofer Compartment
2	12 in. × 14¼ in. × ¾ in.	Sides
2	10½ in. × 14½ in. × ¾ in.	Sides
1	14 in. × 14 in. × ¾ in.	Bottom (Top)
1	12 in. × 12 in. × ¾ in.	Partition
1	10½ in. × 13 in. × ¾ in.	Speaker Board
8 ft. ¾ × ¾ in.		Nailer Cleats

Tweeter Compartment
2 8¾ in. × 12 in. × ½ in.
2 8¾ in. × 11 in. × ½ in.
1 14 in. × 14 in. × ¾ in.

Common to Both Compartments
4 6 in. × 7½ in. × ⅜ in.
2 11 in. × 11 in. × ½ in.
4 ft. of ½ in. material

Components
1 woofer, Norelco 8065/W8 used here
4 mid-range tweeters, 4 in. × 6 in. full range speakers, Norelco 4680-M8

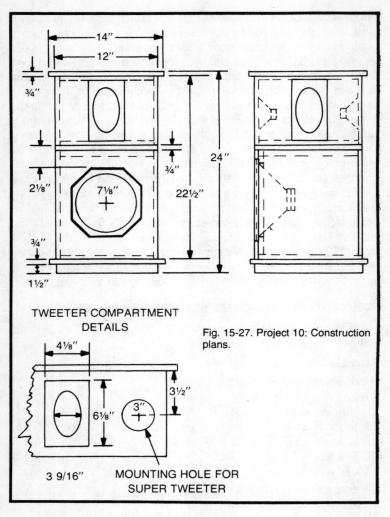

Fig. 15-27. Project 10: Construction plans.

1 Motorola Super Horn Piezo-electric tweeter
1 Capacitor, 12 μf, non-polarized

Directions

The construction plans for this project are shown in Fig. 15-27 and the schematic in Fig. 15-28.

Cut out the parts according to the plans. To make the holes in the tweeter boards fit the speakers, make a heavy cardboard pattern with an oval hole that measures approximately 3-9/16 in. × 5⅝ in. You can make the pattern by trial and error until you get one that fits

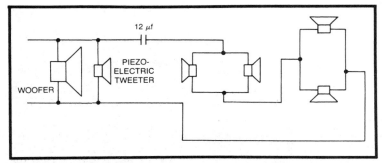

Fig. 15-28. Project 10: Schematic.

snugly over the back of the oval speakers. Use that pattern to mark the outline of the speaker cut-outs on the four boards.

Set the woofer on one of the 12 in. wide sides so that the edge of the woofer is about 2⅛ in. from the upper end of the side. Mark the outline of the woofer's outer edge on the side (Fig. 15-29). Cut the hole slightly larger than the outline to allow the woofer to be centered on its board. The woofer board should fit behind the side with the cut-out for the woofer. There should be no slack between it and the adjacent sides so that it will act as a nailer for those sides. Set it in position behind the side and draw the outline of the woofer cut-out on the speaker board. Center the 7⅛ in. cut-out for the woofer within the marked outline on the board. If you are using a different woofer, the 7⅛ in. diameter may be wrong, and the external cut-out may be

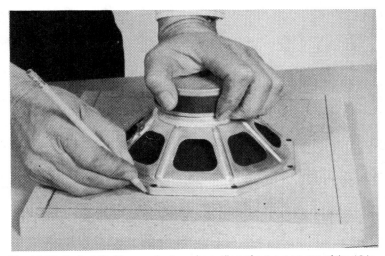

Fig. 15-29. Project 10: Use woofer to make outline of cut-out on one of the 12 in. wide sides of the woofer compartment.

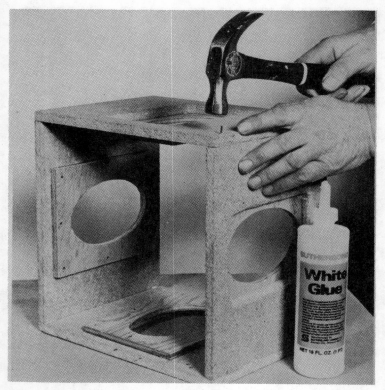

Fig. 15-30. Project 10: Assemble tweeter compartment with nails and glue.

different too. Make a cardboard pattern to fit the back of your woofer to determine the exact diameter of the mounting hole. Locate the speaker bolt mounting holes and drill them. If T-nuts are used, they should be installed on the inside of the speaker board before it is glued and nailed to the external side. Because the Norelco woofers have a butyl rubber surround that extends far enough on the frame to be near the mounting holes, the bolt heads can catch and damage it. Install the bolts with care or, for easier mounting, use silicone rubber glue to install the woofer.

If you plan to use the super-tweeter, it should be installed from outside the enclosure. You should cut the hole for it in a 12 in. wide tweeter box side and set the tweeter temporarily in the hole. Mark the outline of the tweeter on the outside surface of the side and chisel out a 1/8 in. indentation so the tweeter will mount flush with the side.

Glue and nail together the sides of the two sub-enclosure shells (Fig. 15-30). Set each shell over its top and bottom, and notice that the middle partition will be the top for the woofer shell and the bottom for the tweeter section (Fig. 15-31). Mark the outline of the cleat positions; glue and nail the cleats to the lower side of the top, to the upper surface of the bottom, and to the lower side of the partition. Cut a piece of ½ in. plywood or chip board to 11 in. × 11 in. and use it instead of cleats as a nailer on top of the partition. Note that both this nailer and the cleats under the enclosure top in the tweeter compartment should be ½ in. thick, while the cleats in the woofer section are ¾ in. thick.

Drill holes for the speaker leads in the enclosure bottom and in the partition. Mount a screw terminal strip on the bottom with enough lamp cord to reach the partition. Carry the leads through the partition with tight fitting bolts and, for terminals, solder the lugs under the nuts and heads. Do not forget to run the woofer leads

Fig. 15-31. Project 10: Exploded view of enclosure. Note use of thicker material for the cleats in woofer compartment than for those in tweeter compartment.

either from the bottom terminal or from the lower lugs on the partition. These should be long enough to permit the woofer to be wired outside of the enclosure before it is screwed down over its board.

Glue the bottom to the woofer shell, and nail or screw it through the lower sides into the cleats on the bottom. Line the inside of the woofer compartment with a 1 in. to 2 in. thickness of fiberglass. Install the partition with more glue and screws or nails. Then glue and nail the tweeter compartment to the top surface of the partition.

Install a solder lug a short distance from one of the speaker wire bolts in the tweeter compartment. Wire a 12 μf capacitor between the solder lug and the bolt lug. Wire and mount the woofer, using 1 in. × 3/16 in. bolts or silicone rubber glue. Use ½ in. × #8 panhead screws to install the 4 in. × 6 in. speakers. Wire the 4 oval speakers by running a lamp cord directly across the enclosure, between opposing speakers, and with no twists in the cord. This automatically wires the opposing speakers out-of-phase. Connect the two pairs of out-of-phase speakers in series. Apply the test for polarity, Test 13, to find whether or not the opposite speakers move in the same direction when the switch closes. Connect the series lead directly to one side of the speaker line and through the 12 μf capacitor on the other side.

If the piezo-electric tweeter is to be used, mount it over its hole at the rear of the enclosure. Run a piece of lamp cord from the tweeter directly to the conductor bolts. Note that this tweeter is fed directly from the conductors that carry the speaker line from the woofer compartment to the tweeter compartment; the tweeter is not connected through the capacitor.

Fill the upper compartment with loose fiberglass. Test the system to see that all the speakers are working properly; then set the top in place and attach it with screws through the sides into the cleats under the top. Staple a grille cloth around the box, starting at one rear corner and ending there. A 4 ft. section of 48 in. wide grille cloth, split down the middle, will be enough to cover two enclosures. To prevent staining the grille cloth, the trim that covers the edges of the cloth should be finished before it is installed around the enclosure with small brads.

This system gives good sound for the money. You can enhance the omni-directional effect by adding three more tweeters, one to fire out each side, but then you might need to use one of the controls diagrammed in Chapter 2 in order to balance the efficient piezo-electric tweeters with the rest of the system. In this case, the control was not needed because part of the tweeter's output was absorbed by furniture and the walls of the room before the sound reached the listener's ears. A high pass filter to cut off the super-tweeters at 7500 to 10,000 Hz is desirable, but can affect transient response.

PROJECT 11: A MID-RANGE ARRAY

Figure 10-4 shows the project complete and in use.

Parts List
Enclosure
1	11½ in. × 30 in. × ½ to ¾ in.	Speaker Board
2	4 in. × 30 in. × ¾ in.	Sides
2	4 in. × 13 in. × ¾ in.	Top and Bottom
1	2½ in. × 30 in. × ½ in.	Brace

Components
6   4 in. × 6 in. speakers, RCA model used here
1   3 in. tweeter, Westwell AT3AA here
    (these speakers are from McGee Radio Co.)
1   Crossover network, Radio Shack 40-1294 or equivalent
1   Woofer and enclosure (use any woofer and matched enclosure)

Directions

The construction plans for this project are shown in Fig. 15-32, and the schematic in Fig. 15-33.

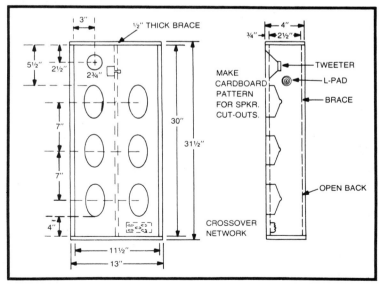

Fig. 15-32. Project 11: Construction plans.

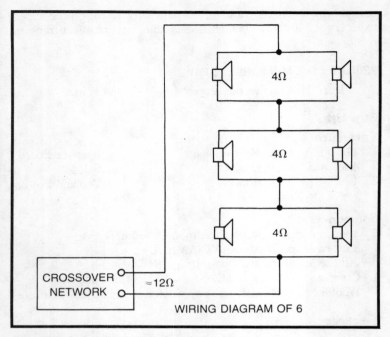

Fig. 15-33. Project 11: Schematic.

The construction of this system is simplicity itself, but do round off the speaker holes at the front edge. This will reduce diffraction by the sharp edge cavity in front of the rear-mounted speakers. The only unusual construction feature is the mid-rib brace down the middle of the speaker board; this brace serves double-duty as a frame for the tweeter control (Fig. 15-34). A cheap tweeter is specified with this project, although some improvement can probably be had by installing a better tweeter. Another trick would be to use a piezo-electric tweeter having a suitable control network, such as is shown in Chapter 2, and a simpler crossover network, such as a choke for the woofer and a capacitor for the mid-range.

PROJECT 12: SINGLE COLUMN STEREO SYSTEM

Figure 10-5 shows this project when completed.

Parts List
Enclosure
1 14 in. × 48 in. × ¾ in.   Front panel
2 12 in. × 48 in. × ¾ in.   Sides
2 12¾ in. × 14 in. × ¾ in.   Top and Bottom
1 12 in. × 12¾ in. × ¾ in.   Partition

Components

12 6 in. × 9 in. speakers, 8Ω impedance.

Directions

The construction plans for this project are shown in Figs. 15-35 and 36 and the schematic in Fig. 15-37.

The speakers in this column can be installed from the rear of the panels, which makes a wrap-around grille cloth feasible. The partition in the middle of the enclosure is put there both to brace the framework and to divide the column into two shorter columns to prevent its acting like a long pipe (Fig. 15-38). Speaker works with proper mates to see that power is divided equally among the 12 units. If you have more than one model, separate the models and put dissimilar speakers together in each sub-branch of

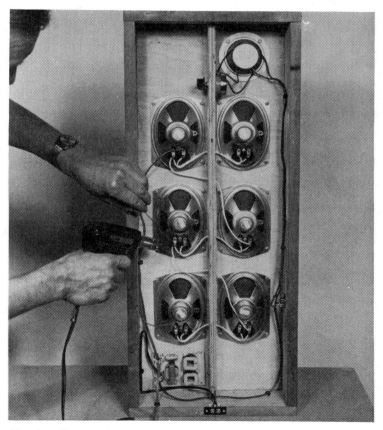

Fig. 15-34: Project 11: Wiring 6-speaker mid-range array. Holes in mid-rib permit connecting wires to run through it and tweeter control to be mounted on it.

Fig. 15-35A. Project 12: Photo of complete shell before installation of speakers.

the circuit. In order to follow the wiring arrangement shown in the plans and yet mix various brands or models of speakers, you must carefully plan the position of each speaker on the column. For example, speakers 1, 2, and 5 are mixed brands. So are numbers 3, 4, and 7; 9, 10, and 6; and 11, 12, and 8.

Wire each parallel trio of speaker mates first, keeping the same polarity for the speakers on the front panel as for those on the side panels. Then wire the two paralleled sets in series, first for the left channel and then for the right channel speakers. Check the final wiring for polarity and, if you have an ohmmeter, measure the dc resistance of each channel. Each channel should have a dc resistance of about 4 Ω, which means that the impedance in the ac circuit will be about 5 Ω to 6 Ω.

Bargain speakers are fine for this system, but do try to find those having a fairly good magnet size, at least 3 to 5 oz., and preferably 8 to 10 oz. Do not get carried away and pay heavily for each speaker; you can easily spend more money that way than you would buying good woofers and tweeters with a wider frequency response.

This system was operated with only a grille cloth for a back. If you like to experiment, you can put on a solid back and vent each compartment. Try an 8 in. × 2 in. simple port. This is the kind of project that leads to experimentation.

PROJECT 13: A SMALL PA COLUMN

This project has a tall, narrow enclosure.

Parts List
Enclosure

2 8⅜ in. × 45 in. × ¾ in.	Sides
2 8⅜ in. × 8 in. × ¾ in.	Top and Bottom
2 6½ in. × 45 in. × ¾ in.	Front and Back
18 ft ¾ in. × ¾ in. pine	Cleats
1 6⅜ in. × 44⅞ in. × ¼ in.	Grille Frame

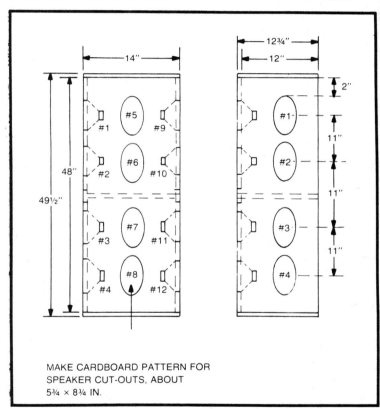

Fig. 15-35B. Project 12: Construction plans of complete shell.

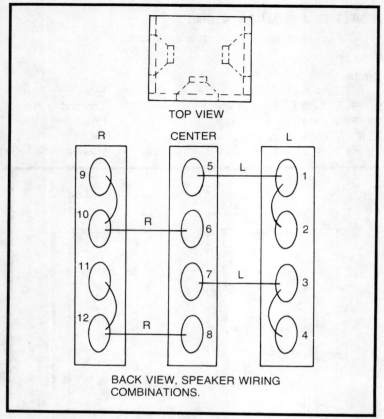

Fig. 15-36. Project 12: Construction plans.

Components

6   4 in. × 6 in. speakers, Norelco 4680/X8.
NOTE: This model is designed especially for PA columns and has a slightly different response as compared to the 4680/M8.

Directions

The construction plans for this project are shown in Fig. 15-39 and the schematic in Fig. 15-40.

These simple columns provide good voice reproduction, but they will not reproduce a low frequency bass range. Below system resonance, at about 170 Hz, the bass response rolls off gradually. However, for good clear voice reproduction, particularly in large rooms where reverberation is a problem, these columns are a good choice.

In the cabinet shown in Fig. 15-41, the speaker board was permanently installed; the speakers were mounted behind the boards, and the backs were screwed on. You can use other methods of construction if you wish. Loosely fill the enclosure with fiberglass before screwing down the back.

The speakers can be wired to match various impedances. These speakers were wired in parallel pairs; the three pairs were then wired in series to give a 12Ω impedance. Two of the columns were wired in parallel and connected to the 4Ω tap on an old tube-type mono public address amplifier. You can experiment with

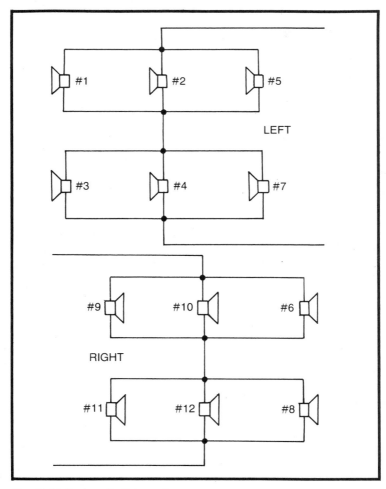

Fig. 15-37. Project 12: Schematic.

Fig. 15-38. Project 12: Rear view. The partition in middle braces the enclosure and divides the column to prevent tuned pipe resonance.

other connections as long as you do not let the total speaker impedance drop below 3 Ω or 4 Ω on solid state amplifiers.

Notice that for greater voice clarity, the speakers specified here are special drivers with extra output in the "presence" range. These speakers are not designed for wide range music reproduction. You could substitute other 4 in. × 6 in. speakers and get satisfactory performance, but if you want a music column speaker, you should go to a different kind of speaker and larger enclosures.

PROJECT 14: A CERAMIC TILE SPEAKER SYSTEM

Figure 11-4 shows this project complete and in use.

Parts List

Enclosure

1   24 in. section of 8 in. flue tile   Shell
2   8½ in. × 8½ in. × ¾ in. plywood   Top and Bottom
1   8½ in. × 8½ in. × ⅜ in. plywood   Top
1   Approximately 7 in. × 7 in. × ¾ in.   Partition
1   10 in. × 10 in./8 mesh hardware cloth screen
1   10 in. × 10 in. grille cloth
3   Rubber feet.
5 ft. of 1½ in. black tape

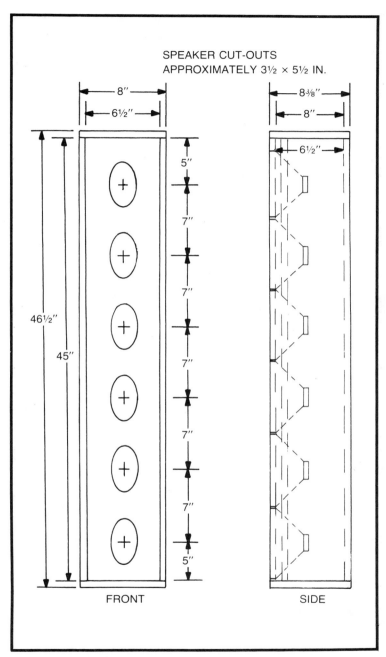

Fig. 15-39. Project 13: Construction plans.

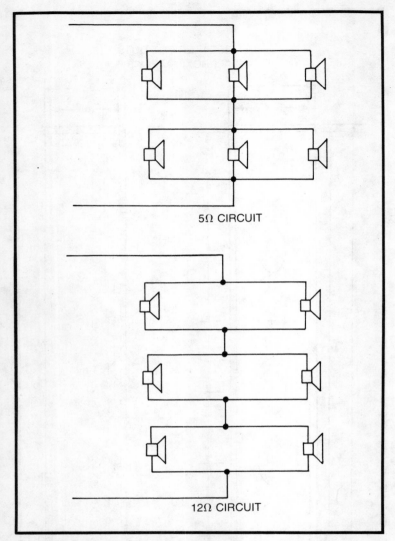

Fig. 15-40. Project 13: Schematic.

Components

1 6 in. extended range speaker, Radio Shack 40-1285
Optional Reflector
1 8 in. × 8 in. piece of ⅛ in. Masonite
1 6 in. × 6 in. piece of glass
1 Stanley 4 in. "T" hinge, screws, 2 small flathead bolts.

Directions

The construction plans for this project are shown in Fig. 15-42.

Set the flue tile on a piece of ¾ in. plywood or chipboard that is large enough to make an end panel, and mark the outline of the tile on the board. Tiles are often irregular; for easy matching after the pieces are cut out, you should identify the side of the board that contacts the tile ends. After marking the end pieces, set the tile on another section of plywood or chipboard, reach down inside the tile, and mark the outline of the partition.

Paint the tile with at least two coats of flat wall paint. If you prefer a more durable and more easily cleaned surface, you can use enamel, but a glossy enamel will exaggerate any imperfections in the outer surface of the tile.

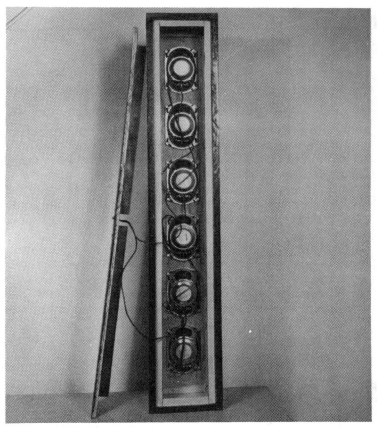

Fig. 15-41. Project 13: Interior view. Notice stiffener on back. The enclosure should be loosely filled with damping material before screwing down back.

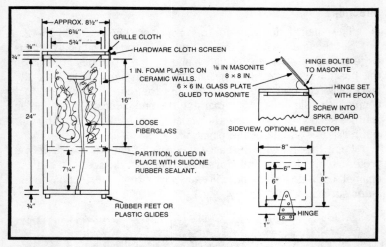

Fig. 15-42. Proejct 14: Construction plans.

While the paint is drying, you can cut out the end pieces and the partition. Use the speaker board as a pattern to cut a doughnut-shaped piece of ⅜ in. plywood to fit around the speaker (Fig. 15-43). Make the cut-outs for the speaker both in the speaker board and in the ⅜ in. ring according to the plans, and glue and nail the ring to the speaker board. Install speaker wire in the bottom panel, and drill a hole through the center of the partition to allow the wire to pass through it. This wire should be long enough to reach out of the top of the enclosure to make soldering to the speaker terminals easier when you later mount the speaker. Paint the top surface, the edges of the speaker board assembly, and the edges of the bottom panel all flat black.

Pass the speaker cord through the hole in the partition and insert the partition into the tile. Position it at the proper level and then move it slightly out of position. Coat the walls of the tile with silicone rubber glue where the partition will contact them. Move the partition back into place. Seal the joints top and bottom around the partition and the tile with silicone rubber sealant.

Glue the bottom panel to the tile with silicone rubber sealant. Pull the speaker cord up through the partition to take up the slack in the lower compartment, and seal the hole around the cord in the partition. Use rubber cement or contact cement to glue a 1 in. layer of foam plastic to the inner walls of the tile in the speaker compartment. Cover the upper surface of the partition with a 2 in. or 3 in. layer of fiberglass. Glue the prepared speaker board to the top edge of the tile.

When the glue has set, you can place the speaker face down over the hole in the top panel and test it. If the system resonance is higher than desirable, you can add fiberglass or Dacron filler to the speaker compartment. If you do not have the facilities for testing, install the speaker by soldering the wires to the terminals on the back of the speaker, and then front mount the speaker with panhead sheet metal screws. If performance is satisfactory, remove the screws, run a bead of silicone rubber under the speaker rim, and screw it down again.

Cover the up-turned speaker with 8-mesh hardware cloth painted flat black. The 8-mesh cloth, which has a strand of wire every 1/8 in. in a square cross pattern, will protect the speaker from sharp objects that might drop on it and puncture the cone. Place grille cloth over the wire cloth and staple it to the edges of the speaker board, stretching it taut before you drive in each staple. Cover the staples with black tape. To make the bottom panel edges match the top, you can use the same tape on them.

You can make an optional reflector which will prevent loss of high frequency response to the ceiling. In listening tests, some people preferred the reflector; others liked the more diffuse effect of the simple up-turned speaker.

To make the reflector, bolt the hinge to the 8 in. × 8 in. Masonite as shown in the drawings. Use a small flathead bolts and countersink the heads in the rough side of the Masonite. The hinge fastens to the opposite smooth side. Paint the assembly flat black.

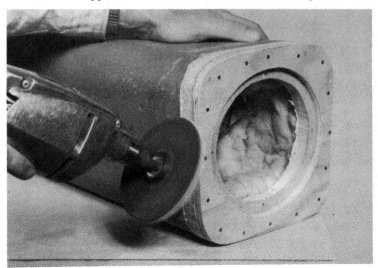

Fig. 15-43. Project 14: Sand edges of speaker board assembly to make it conform to the shape of the ceramic tile.

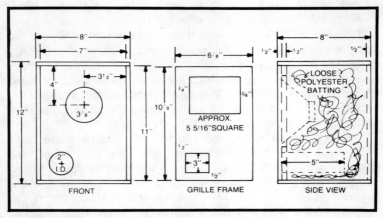

Fig. 15-44. Project 15: Construction plans.

Use silicone rubber sealant such as GE auto seal to glue the glass plate to the rough side of the Masonite. Fasten the reflector to the enclosure by screws through the hinge base into the speaker board.

You can experiment with various reflection angles, but for small rooms, an angle of 50 to 55 degrees is about right. When you have found the optimum angle, prop the reflector at that angle and apply epoxy glue to the hinge to hold it there.

PROJECT 15: A COMPACT STUFFED BOX SPEAKER

This project is shown complete and in use in Fig. 11-5.

Parts List
Enclosure
2   8 in. × 11 in. × ½ in.   Sides
2   8 in. × 8 in. × ½ in.   Top and Bottom
2   7 in. × 11 in. × ½ in.   Front and Back
1   6⅞ in. × 10⅞ in. × ½ in.   Grille Frame
2   Frozen orange juice cans, taped and damped. Duct

Components
5 in. high compliance speaker, Radio Shack 40-1292

Directions
The construction plans for this project are shown in Fig. 15-44.

You can build this enclosure with glue and nails; no cleats are necessary if you seal the joints with caulking material.

Prepare the tuning duct by taping together two sections of frozen orange juice cans. These ducts can be damped by coating them with asphalt roofing cement and more tape. If you plan to tune the box yourself, don't use the asphalt cement: it will scrape off on the front panel when you try to insert the duct from outside the completed enclosure. Instead, use masking tape to brace the duct; then coat the masking tape with white glue.

Cut out the parts and mark the location of the front panel on the sides, top, and bottom, leaving space at the front for the grille frame. Glue and nail a side to the front panel. Next, glue and nail down the bottom panel, making a butt joint with the previously installed side. Overlap the other side of the front panel enough to cover the edge of the next side. Install the next side, then the top. After the glue has dried, caulk all the inside joints.

Pass the speaker wires through the back by an approved leak-tight method and install the back with glue and nails. When the glue has set, you can caulk the joints between the back and the sides by reaching through the speaker hole with a long caulking tool, such as a narrow-bladed screwdriver.

Loosely fill the box with Dacron, but leave the area around the rear duct opening clear. Apply veneer to the outside of the enclosure (Fig. 15-45) and mount the speaker. Staple the grille cloth to the grille frame and press it in for a friction fit or, if necessary, use Velcro or tape to hold it in place.

Note: In case you missed the discussion about leaky speakers in Chapter 11, you should test these speakers on a 0.1 ft.^{3} test box for air leaks. A 90 Hz resonance, for example, should be raised to about 135 Hz or higher on the test box. If the resonance changes very little, check the surround fabric near the outer edges for inadequate

Fig. 15-45. Project 15: Fill enclosure with Dacron, then hollow out a tunnel where port goes. The single orange juice can here is only part of the final tuning duct.

311

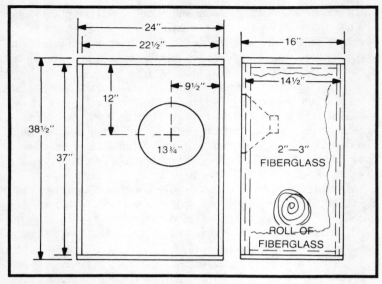

Fig. 15-46. Project 16: Construction plans.

sealing. Go back to Chapter 11 to see how to seal leaky cloth suspensions.

PROJECT 16: LOW COST SUB-WOOFER

This project has a very simple enclosure.

Parts List

Enclosure

2 22½ in. × 37 in. × ¾ in.	Front and Back
2 16 in. × 24 in. × ¾ in.	Top and Bottom
2 16 in. × 37 × ¾ in.	Sides
1 2 by 4 × approx 33 in.	Back brace
25 ft. of 1 by 2 pine	Cleats and glue blocks
16 ft. of 16 × 2 to 3 in. fiber glass batting	Damping material

Components

15 in. woofer, high compliance, Stk. #CTS-15-10-8, McGee Radio Co.

Directions

The construction plans for this project are shown in Fig. 15-46.

The construction of the box is conventional except for a 1 in. × 4 in. brace on the speaker board, as shown in Fig. 15-47, and a 2 in. × 4 in. brace on the back. Use plenty of glue at each matching surface. If you use nails to put this box together, use the screw-thread nails that clinch tighter than ordinary nails. They will hold the parts together under pressure while the glue sets.

You can install the woofer from the back of the speaker board, so make the back removable to allow access to the speaker if necessary. Put a strip of foam weatherstripping around the back, just inside the screw holes, where it will seal the space between the back and the cleats.

A large object looks smaller if it is painted a plain dark color, so it is a good idea to paint the exterior flat black, or cover it with black grille cloth such as black decorator burlap.

That's all there is to this essentially simple project, except for a conventional crossover network which you can improvise either according to the directions in Chapter 2 or, if you plan to use this as a single woofer to carry the mixed bass of a stereo system, to the plans shown in Chapter 11.

PROJECT 17: A LOW COST WOOFER-TWEETER SYSTEM

Figure 14-5 shows this project when completed.

Parts List
Enclosure
2   8 in. × 10 in. × ½ in.   Sides
2   8 in. × 6½ in. × ½ in.   Top and Bottom

Fig. 15-47. Project 16: Interior view, showing use of 2 in. × 4 in. braces on back and speaker board. This system requires a considerable amount of damping material.

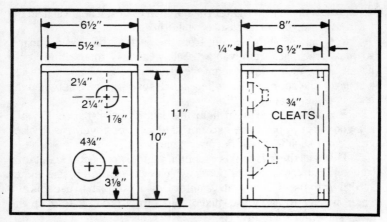

Fig. 15-48. Project 17: Construction plans.

2   5½ in. × 10 in. × ½ in.   Front and Back
3 ft. ¾ in. × ¾ in. Pine cleat material.
1   5¼ or 5½ in. × 10 in. car speaker grille. (K-Mart SG-169 or equivalent)
¾ in. Cloth Tape,   Mystic

Components
1   5¼ in. woofer
1   2½ tweeter. (these speakers from McGee Radio, Stock # 525-TW-2)
1   4 µf non-polarized capacitor
1   10Ω resistor
1   5Ω resistor
Resistors can be 2 to 5 W.

Directions

The construction plans for this project are shown in Fig. 15-48 and the schematics in Fig. 15-49.

This enclosure was built around an inexpensive 5¼ in. × 10 in. plastic grille sold for 6 in. × 9 in. car speakers (Fig. 15-50). If you do not use this grille, you can make changes in dimensions, but try to keep the same cubic volume. One change you would have to make would be to increase the depth that the speaker board is set back from the front edge of the box. Here it is set back only ¼ in. to accommodate the thin grille. Almost any homemade grille frame will be too thick for the slight space available, so you should increase the depth of the enclosure from 8 in. to 8¼ in. to allow a full ½ in. space for the grille.

Although the plans show ½ in. material for all parts, you can use ⅜ in. thick chipboard for the sides and back. In order to use the

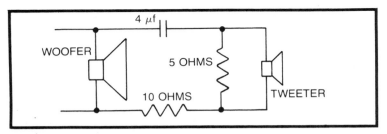

Fig. 15-49. Project 17: Schematic.

special grille, the speakers are installed in back of the speaker board and the removable back is screwed into ¾ in. cleats. Again, if you use a homemade grille, you can use front mounting and seal the entire box.

After building the box, prepare 8 pieces of ¾ in. wide Mystic cloth tape 1⅜ in. long, which will be about 1.5 g. of added mass. Attach the tape to the woofer cone in a radial pattern, extending from near the center of the cone out to the edge of the piston area.

Install the speakers and wire them. If the tweeter's terminals are not marked for polarity, you can assume that the left terminal of the woofer should be connected to the left terminal of the tweeter. Wire a 4 μf capacitor between the speaker terminal lugs on one side and a 10Ω resistor on the other side. Connect a 5Ω resistor across the tweeter terminals. Try the system and, if the highs are in proper balance, fill the enclosure with loose damping material and screw down the back.

Fig. 15-50. Project 17: All the parts assembled.

315

16

On Your Own

The plans in Chapter 15 show projects that have been built and tested. Here are some plans that should work out well, but have not actually been tried.

Figures 16-1 and 16-5 show enclosure plans, a schematic diagram, and other information on equalized bass reflex systems. The same equalizer was discussed in Chapter 7, but here again are the important points on its use.

Set the system cut-off frequency (f_{AUX}) by this formula:

$$C_1 = C_2 = 23.2/f_{aux}$$

For the speaker system shown in Fig. 16-1, the values of C_2 and C_3 should be about 0.93. For the one in Fig. 16-2, C_2 and C_3 should be about 0.83 μf.

The variable resistance labeled R_4 permits you to vary the boost up to 10 dB. To get the theoretically right amount of boost (6 dB), substitute a 30Ω resistor for R_4, a 970Ω resistor for R_6, and connect the lead from R_2 to the junction of R_4 and R_6.

To try the equalizer with other speakers, use these formulas to get the right box volume and tuning:

$$V_B = 4.1 \ (Q^2) \ V_{as}$$
$$f_B = (0.3) \ (f_s/Q) \text{ (after Keele)}$$

Again: do not build the speaker systems outlined in Fig. 16-1 and 16-2 unless you intend to build an equalizer to go with them. With higher tuning, you can redesign the boxes for unequalized reflex operation, but a rumble filter is desirable.

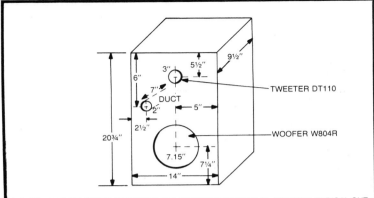

Fig. 16-1. Suggested enclosure for Speakerlab kit, S1S, converted to equalized bass reflex system.

If you are planning to build the equalizer, send 25 cents to Speakerlab, whose address is listed in the appendix, for their "Bass Reflex Equalizer" bulletin and study it. For easy construction, you should also order their printed circuit card.

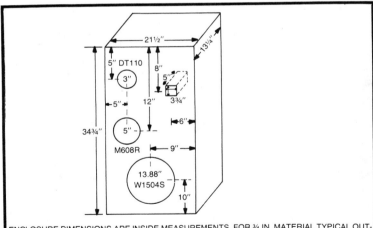

Fig. 16-2. Suggested enclosure for Speakerlab woofer, W1504S, with mid-range driver and tweeter.

317

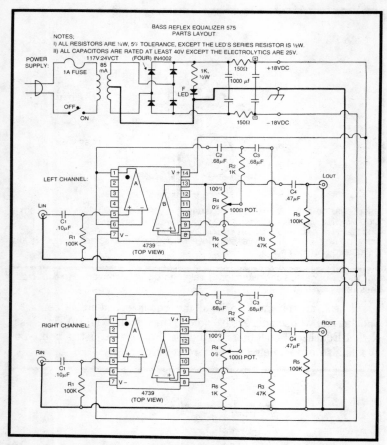

Fig. 16-3. Schematic diagram of Speakerlab bass reflex equalizer.

The woofer system shown in Fig. 16-6 was designed to fit as a base under the mid-range/tweeter array described as Project 11. This uses the same heavy Heppner 6 in. × 9 in. speaker that was tested with the array in a slightly smaller, side-firing box of similar but not identical design. Tests with the system indicated that most listeners preferred a front-firing woofer, but the original enclosure matched the width of the mid-range enclosure, which was desirable for good appearance, only when the woofer was in one side.

The oval woofer was chosen to permit an enclosure height of 12 in. This height puts the mid-range array at a reasonable level for seated listeners when it is perched on the woofer enclosure. The narrow duct was dictated by available space, so check the tuning, if

PARTS LIST FOR SPEAKERLAB BASS REFLEX EQUALIZER 575

1—CIRCUIT CARD (OPTIONAL).
1—117 V PRIMARY, 24 OR 25 VCT SECONDARY, 85 mA (OR MORE) TRANSFORMER.
4—1N4002 OR OTHER 100 V, 1A (OR GREATER) RECTIFYING DIODES.
4—1000 µf, 25 V OR GREATER ELECTROLYTIC CAPACITORS.
1—LED (LIGHT EMITTING DIODE).
2—RAYTHEON 4739 INTEGRATED CIRCUIT OPERATIONAL AMPLIFIERS.
2—14-PIN DUAL-IN-LINE IC SOCKETS (OPTIONAL).
2—100Ω TRIM POTS FOR MOUNTING DIRECTLY ON PRINTED CIRCUIT BOARD (ALTERNATIVELY YOU CAN USE ONE 100 OHM DUAL-GANGED, ¼ W OR GREATER, POT FOR PANEL-MOUNTING).
2—150Ω, 5%, ¼W CARBON FILM RESISTOR (COLOR CODE: BROWN-GREEN-BROWN-GOLD).
4—1 K, 5%, ½ W CARBON COMPOSITION RESISTOR (BROWN-BLACK-RED-GOLD)—GOES NEXT TO LED.
2—47 K, 5%, ¼ W CARBON FILM RESISTOR (YELLOW-PURPLE-ORANGE-GOLD).
4—100 K, 5%. ¼ W CARBON FILM RESISTOR (BROWN-BLACK-YELLOW-GOLD).
2—.10µf, 25 V OR GREATER CAPACITOR.
2—.47 µf, 25 V OR GREATER CAPACITOR.
4—.68 µf, 25 V OR GREATER CAPACITOR.
4—PHONO JACKS FOR INPUTS AND OUTPUTS.
1—ON-OFF SWITCH.
1—FUSEHOLDER AND 1 A FUSE.
1—LINE CORD AND PLUG.

Fig. 16-4. Parts list for Speakerlab bass reflex equalizer 575.

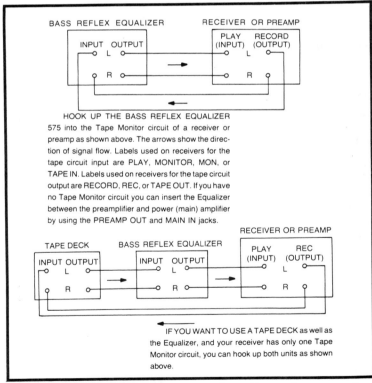

HOOK UP THE BASS REFLEX EQUALIZER 575 into the Tape Monitor circuit of a receiver or preamp as shown above. The arrows show the direction of signal flow. Labels used on receivers for the tape circuit input are PLAY, MONITOR, MON, or TAPE IN. Labels used on receivers for the tape circuit output are RECORD, REC, or TAPE OUT. If you have no Tape Monitor circuit you can insert the Equalizer between the preamplifier and power (main) amplifier by using the PREAMP OUT and MAIN IN jacks.

IF YOU WANT TO USE A TAPE DECK as well as the Equalizer, and your receiver has only one Tape Monitor circuit, you can hook up both units as shown above.

Fig. 16-5. How to hook up the bass reflex equalizer with a receiver or pre-amp.

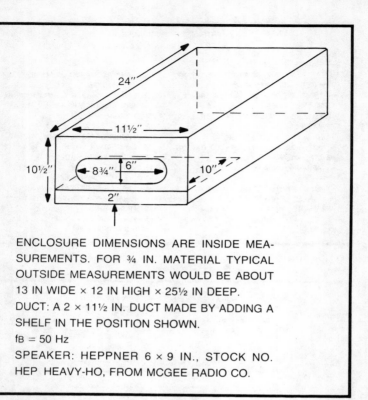

Fig. 16-6. Suggested woofer enclosure for use with Project 11, the mid-range array.

possible. There is a chance that it should be slightly longer than the 10 in. length shown. Keep the duct clear and smooth.

So we have come to the end of this book, not with a neatly finished project, but with some suggestions for future experiments. After all, getting you involved in designing your own speaker system was the aim of this book.

Appendix I
Useful Formulas

Electrical Units Used in Speaker Circuits

Inductance

The unit of inductance is the henry

All smaller units should be converted to henrys before using the number of formulas.

$$1 \text{ millihenry (mH)} = 0.001\text{H or } 1 \times 10^{-3}\text{H}$$
$$1 \text{ microhenry } (\mu\text{H}) = 0.000001\text{H or } 1 \times 10^{-6}\text{H}$$

Capacitance

The unit of capacitance is the farad

All smaller units should be converted to farads before using the numbers in formulas.

$$1 \text{ microfarad } (\mu\text{f}) = 0.000001 \text{ farad or } 1 \times 10^{-6} \text{ farad}$$
$$1 \text{ picofarad (pf)} = 0.000000000001 \text{ farad or } 1 \times 10^{-12} \text{ farad}$$

Ohm's Law for Speakers

For alternating current, the Ohms's Law formulas are:

$$E = IZ$$
$$I = E/Z$$
$$Z = E/I$$

E is the voltage in volts
I is the current in amperes
Z is the impedance in ohms.

321

Worked Examples:

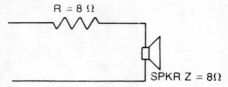

An 8Ω resistor is put in series with an 8Ω tweeter.
How much is the sound level reduced?
Assume any drive voltage that is convenient; for example, 16 V. The nature of the speaker's impedance depends on the frequency, so we will assume that it is purely resistive.
Then:

$$I = 16/16 = 1 \text{ A}$$

The voltage across the speaker is:

$$E = 1 \times 8 = 8 \text{ V}$$

Before adding the resistor, the voltage across the speaker was 16 V. So the ratio of voltage in the new circuit to the old one is:

$$\frac{\text{New voltage}}{\text{Old voltage}} = \frac{8}{16} = 0.5$$

The dB/ voltage ratio chart tells us that for a voltage ratio of 0.5, the output is down 6 dB.

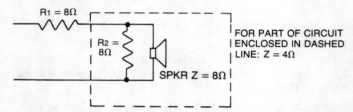

In the same tweeter circuit, an additional resistor is added parallel to the tweeter. What is the tweeter output now compared to its output with no resistors in its circuit? First, get the impedance of the tweeter and its parallel resistor. Since the paralleled resistances are equal, then the net impedance of the tweeter and paralleled resistor branch is 8/2 or 4Ω.

The total impedance in the circuit is 8 + 4 = 12Ω.
And:

$$I = 16/12 = 1.33 \text{ A}$$

The voltage across the tweeter and its parallel resistor branch is:

$$E = 1.33 \times 4 = 5.32 \text{ V}$$

And the tweeter's output is:
$$\frac{\text{New voltage}}{\text{Old voltage}} = \frac{5.32}{16} = 0.33$$

From the dB chart, the output is down 10 dB.

Reactance & Resonance
Capacitors
Capacitive reactance:
$$X_c = \frac{1}{2\pi f C}$$

To get the value of a capacitor that will give a specified reactance at a given frequency:
$$C = \frac{1}{2\pi f X_c}$$

Inductance
Inductive reactance:
$$X_L = = 2\pi f L$$

To get the value of a choke that will give a specified reactance at a given frequency:
$$L = \frac{X_L}{2\pi f}$$

Resonance
$$f = \frac{1}{2\pi \sqrt{LC}}$$

Worked example for resonance:
A 1.875 mH choke is wired into a peak suppressor filter circuit with a 7.5 μf capacitor. What frequency will be most affected?
$$f = \frac{1}{6.28 \quad \sqrt{0.001875 \times 0.0000075}} = 1342 \text{ Hz}$$

Notice that in the formula for resonance, it is the *product* of L × C that determines the frequency of resonance. You can increase L and decrease C by the proper amounts and still have the same frequency. This permits you to make filters that are effective at the frequency you want with various values of components.

For parallel resistors:

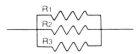

Where $R_1 = R_2 = R_3$, and N is the number of resistors
$$\text{Total } (R_T) = R/N$$
For mixed value resistors:
$$R_T = \frac{1}{1/R_1 + 1/R_2 + 1/R_3}$$
If n = 2
$$R_T = \frac{R_1 \times R_2}{R_1 + R_2}$$

For series resistors:

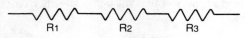

$$R_T = R_1 + R_2 + R_3$$

For parallel capacitors:

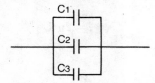

$$C_T = C_1 + C_2 + C_3$$

For series capacitors:

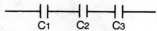

Where $C_1 = C_2 = C_3$ and N is the number of capacitors.
$$C_T = C/N$$
For mixed value capacitors:
$$C_T = \frac{1}{1/C_1 + 1/C_2 + 1/C_3}$$
If N = 2:
$$C_T = \frac{C_1 \times C_2}{C_1 + C_2}$$

For parallel chokes:

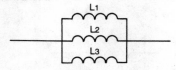

Where $L_1 = L_2 = L_3$, and N is the number of chokes.
$$L_T = L/N$$
For mixed value chokes:
$$L_T = \frac{1}{1/L_1 + 1/L_2 + 1/L_3}$$
If $N = 2$:
$$L_T = \frac{L_1 \times L_2}{L_1 + L_2}$$

For series chokes:

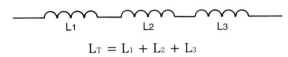

$$L_T = L_1 + L_2 + L_3$$

For resistor and capacitor in series:

$$Z = \sqrt{R^2 + X_c^2}$$

Worked example:
Let $R = 10\ \Omega$, $C = 0.000002$ or $2\ \mu f$.
What is the impedance at 10,000 Hz?

$$X_c = \frac{1}{2\pi fC}$$

$$= \frac{1}{(6.28)\ (10,000)\ (0.000002)} = 8\Omega$$

And:

$$Z = \sqrt{100 + 64} = 12.8\ \Omega$$

For resistor and capacitor in parallel:

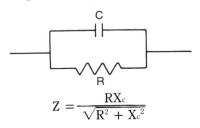

$$Z = \frac{RX_c}{\sqrt{R^2 + X_c^2}}$$

Worked example:
> Let R = 10Ω, C = 0.000002 f or 2 μf.
> What is the impedance at 10,000 Hz?
> From the example above: $X_c = 8Ω$ at 10,000 Hz.
> So:
>
> $$Z = \frac{10 \times 8}{\sqrt{100 + 64}} = \frac{80}{12.8} = 6.25 \, Ω$$

For resistor and choke in series:

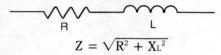

$$Z = \sqrt{R^2 + X_L^2}$$

Worked example:
> Let R = 10Ω, L = 0.000127 H or 0.127 mH
> What is the impedance at 10,000 Hz?
>
> $$X_L = 2\pi fL = 6.28 \times 10,000 \times 0.000127 = 8Ω$$
>
> So:
>
> $$Z = \sqrt{100 + 64} = 12.8Ω$$

For resistor and choke in parallel:

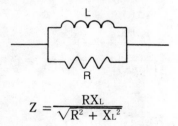

$$Z = \frac{RX_L}{\sqrt{R^2 + X_L^2}}$$

Worked example:
> Let R = 10Ω, L = 0.000127 H or 0.127 mH
> What is the impedance at 10,000 HZ?
> From the example above:
>
> $$X_L = 8Ω \text{ at } 10,000 \text{ Hz.}$$
>
> So:
>
> $$Z = \frac{10 \times 8}{\sqrt{100 + 64}} = \frac{80}{12.8} = 6.25Ω$$

For a resistor, choke, and capacitor in parallel:

$$Z = \frac{R\,X_L X_c}{\sqrt{X_L^2 X_c^2 + R^2(X_L - X_c)^2}}$$

Worked example:

Let R = 10Ω, L = 0.00159 or 1.59 mH, and C = 0.000016 or 16 μf.

What is the impedance at 1000 Hz?

$$X_L = 2\pi f L = (6.28) \times 1000 \times 0.0015 \approx 10\,\Omega$$

$$X_c = \frac{1}{2\pi f C}$$

$$= \frac{1}{6.28 \times 1000 \times 0.000016} \approx 10\,\Omega$$

$$Z = \frac{10 \times 10 \times 10}{\sqrt{100 \times 100 + 100\,(10 - 10)^2}}$$

$$= \frac{1000}{100.5} \approx 10\,\Omega$$

What is the impedance at 500 Hz?

At 500 Hz the reactance of L will be halved, compared to 1000 Hz, and that of C will be doubled.

So: $X_L = 5\,\Omega$ and $X_c = 20\,\Omega$ at 500 Hz.

$$Z = \frac{10 \times 5 \times 20}{\sqrt{25 \times 400 + 100\,(5 - 20)^2}}$$

$$= \frac{1000}{\sqrt{32500}} = \frac{1000}{180.3} = 5.55\,\Omega$$

Conversion Factors for Speaker Compliance

In this book speaker compliance is measured in cm/dyne- the CGS (centimeter-gram-second) system because the centimeter is a convenient unit to use in measuring cone diameter. Sometimes compliance is shown in mixed units of mm/Newton or even as its reciprocal term—stiffness—in Newton/Meter, the MKS (meter-kilogram-second) system.

To change cm/dyne to mm/Newton:

$$1 \text{ cm} = 10 \text{ mm}$$
$$1 \times 10^5 \text{ dynes} = 1 \text{ Newton}$$

So:

$$\frac{1 \text{ cm}}{\text{dyne}} \times \frac{10 \text{ mm}}{\text{cm}} \times \frac{1 \times 10^5 \text{ dynes}}{\text{Newton}}$$

cancels out to:

$$1 \times 10 \text{ mm} \times \frac{1 \times 10^5}{\text{Newton}} = 1 \times 10^6 \text{ mm/Newton}$$

So:

$$C_{ms} \text{ (cm/dynes)} \times 1 \times 10^6 = C_{ms} \text{ (mm/N)}.$$

To change C_{ms} in cm/dynes to stiffness (S) in Newton/Meter, first change C_{mx} in cm/dynes to S in dynes/cm by:

$$S = 1/C_{ms}$$

And:

$$\frac{1 \text{ dyne}}{\text{cm}} \times \frac{1 \text{ Newton}}{1 \times 10^5 \text{ dynes}} \times \frac{1 \times 10^2 \text{ cm}}{1 \text{ meter}}$$

cancels out to:

$$1 \times \frac{1 \text{ Newton}}{1 \times 10^5} \times \frac{1 \times 10^2}{1 \text{ meter}} = 1 \times 10^{-3} \text{ Newton/meter}$$

So: S in dynes/cm $\times 1 \times 10^{-3}$ = S (Newton/Meter)

Worked example:

The compliance of 12 in. speaker is listed as 0.416×10^{-6} cm/dynes.

What is the stiffness in N/m?

$$S \text{ (dynes/cm)} = \frac{1}{0.417 \times 10^{-6}} = 2.4 \times 10^6 \text{ dynes/cm}$$

$$S \text{ (N/m)} = 2.4 \times 10^6 \text{ dynes/cm} \times 1 \times 10^{-3}$$
$$= 2.4 \times 10^3 \text{ N/m}$$
$$= 2400 \text{ N/m}$$

Speaker Formulas for Constant Voltage Sound Distribution Systems

Power formulas:

$$P = \frac{E^2}{Z} \qquad Z = \frac{E^2}{P}$$

where:

P = the power in watts
E = the voltage in volts
Z = the impedance in ohms

APPROXIMATE VOLTAGE RATIO (+)	dB	APPROXIMATE VOLTAGE RATIO (−)
1	+0−	1
1.1	1	0.9
1.25	2	0.8
1.4	3	0.7
1.6	4	0.6
1.8	5	0.55
2	6	0.5
2.25	7	0.45
2.5	8	0.4
2.8	9	0.35
3.2	10	0.32
3.5	11	0.28
4	12	0.25
5.6	15	0.18
8	18	0.13
16	24	0.06

TO CALCULATE ANY VOLTAGE RATIO:
$dB = 20 \log E_1/E_2$

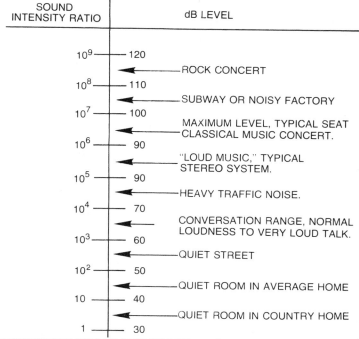

SOUND INTENSITY RATIO	dB LEVEL	
10^9	120	
		ROCK CONCERT
10^8	110	
		SUBWAY OR NOISY FACTORY
10^7	100	
		MAXIMUM LEVEL, TYPICAL SEAT CLASSICAL MUSIC CONCERT.
10^6	90	
		"LOUD MUSIC," TYPICAL STEREO SYSTEM.
10^5	90	
		HEAVY TRAFFIC NOISE.
10^4	70	
		CONVERSATION RANGE, NORMAL LOUDNESS TO VERY LOUD TALK.
10^3	60	
		QUIET STREET
10^2	50	
		QUIET ROOM IN AVERAGE HOME
10	40	
		QUIET ROOM IN COUNTRY HOME
1	30	

COMPARISON OF THE LOUDNESS OF VARIOUS SOUNDS BY dB LEVEL AND SOUND INTENSITY RATIO.

To Choose impedance matching transformers:

1. Decide how much power is to be available to each speaker by:
   a. cubic volume and acoustical character of each location
   b. desired sound level at each location
   c. available amplifier power—make sure total isn't exceeded.
2. Substitute the transformer primary impedance for Z and the voltage of distribution system for E in the formulas above to find the right impedance transformer.

Worked example: A 50W amplifier is used in a 70V line. What impedance transformer is needed where the desired power to the speaker is 10W.

$$Z \text{ (of transformer)} = \frac{(70)^2}{10} = \frac{5000}{10} = 500\Omega$$

The needed transformer should have a primary impedance of 500Ω and a secondary impedance to match that of the speaker. No more than five such speakers can be used with this amplifier.

To choose the correct constant voltage transformer tap:

1. Choose a transformer with a matching secondary impedance—an 8 Ω secondary for an 8Ω speaker.
2. Connect the desired power tap—10W for above example—to speaker.
3. Connect the constant voltage distribution line to the primary.

Appendix II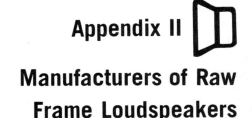

Manufacturers of Raw Frame Loudspeakers

Altec-Lansing
1515 South Manchester Ave.
Anaheim, CA 92803

Manufactures a line of full range speakers including duplex models and separate woofers, high frequency drivers, and crossover networks. Publishes a book, *Loudspeaker Enclosures: Their Design and Use,* available for $2.00. The book contains general information on loudspeaker enclosures, tuning instructions for bass reflex boxes, and several outer dimensional drawings of suggested enclosures for Altec speakers.

Electro-Voice, Inc.
600 Cecil St.
Buchanan, MI. 49107

Manufactures a line of extended range speakers, full range speakers, and separate woofers, high frequency horns and drivers, and crossover networks. Gives full Thiele/Small specifications on woofers and full range speakers. Publishes full data sheets for each speaker, plus builders' plans for various sizes of bass boxes.

JBL
8500 Balboa Blvd.
Northridge, CA 91329

Manufactures a line of full range, extended range, and low frequency speakers, passive radiators, mid-range drivers, high frequency drivers and horns, plus crossover networks. Enclosure construction kit available at $5, but only from JBL dealers who carry JBL raw frame speakers, or by mail from JBL Technical Information Department. This kit includes a brochure on bass reflex techniques and a set of construction drawings for enclosures.

Motorola, Inc.
Piezoceramic Products Group
2553 Edgington
Franklin Park, IL 60131

Manufactures several models of piezo-electric tweeters.

Speakerlab, Dept. BRE
5500—35th NE
Seattle, WA 98105

Manufactures some of the line of about thirty full range, woofers, mid-ranges, tweeters and horns they sell. Publishes complete Thiele/Small specifications in catalog which also has considerable useful information on speakers. Will send catalog free from above address. Also publishes a number of low-cost pamphlets—such as the 25 cent "Bass Reflex Equalizer"— which are listed in the catalog.

Mail-Order houses that offer speakers and components

Burstein-Applebee
3199 Mercier St.
Kansas City, MO 64111

Carries a few full range of speakers plus an assortment of woofers and tweeters in back of catalog. Catalog price: $2.00.

McGee Radio Co.
1901 McGee St.
Kansas City, MO 64108

Carries an extremely wide variety of woofers, mid-range drivers and tweeters plus passive radiators. Some crossover networks and components. Catalog free from above address.

Olson Electronics
260 S. Forge St.
Akron, OH 44327

Carries a line of house brand speakers. Catalog free.

Speakerlab, Dept. BRE
550—35th NE
Seattle, WA 98105

Carries a line of about thirty full range, woofers, mid-range drivers, tweeters and horns, along with crossover components such as high power L-pads that will work in systems up to 200 W RMS per channel. Catalog has considerable useful information on speakers, including complete specifications. Free on request from above address.

SOME NAME BRAND RAW FRAME SPEAKERS FROM 10" TO 15" IN DIAMETER

MANUFACTURER	MODEL NO.	SIZE & KIND	fS (Hz)	Q	V(as) (FT.)	REC. VB (ft.³)	TYPE	REC. TWEETER & CROSSOVER
ALTEC LANSING	601-8D	12 IN 2-WAY COAXIAL	39			3.5	CLOSED BOX	
ALTEC LANSING	604#-8G	15 IN 2-WAY COAXIAL	25			9	CLOSED BOX	
ELECTRO-VOICE	MC12A	12 IN SINGLE CONE	50	0.78	5.1	6.0	PORTED	HF1 HORN & Xover AT 3500 Hz
ELECTRO-VOICE	SP12C	12 IN. SINGLE CONE	45	0.67	5.9	5.5	PORTED	BB1 HORN & XOVER AT 3500 Hz
ELECTRO-VOICE	12TRXC	12 IN. COAXIAL	50			7	PORTED	BB1 HORN & XOVER AT 3500 Hz
ELECTRO-VOICE	SP15A	15 IN. SINGLE CONE	40	0.4	9.9	7.5	PORTED	
JBL	LE10A	10 IN. LOW FREQUENCY				1 TO 4	PORTED	LE 20, LX11 XOVER AT 2500 Hz
JBL	LE12C	12 IN. FULL RANGE 2-WAY				1.5 TO 6.0	PORTED	
JBL	LE-14C	14 IN FULL RANGE 2-WAY				1.5 TO 6.0	PORTED	075, N2400 XOVER AT 2500 Hz
JBL	D130	15 IN. EXTENDED RANGE				4 TO 12	PORTED	

THIS LIST INCLUDES ONLY A FEW OF THE MANY MODELS AVAILABLE FROM THESE MANUFACTURERS. FOR EXAMPLE, ELECTRO-VOICE OFFERS ADDITIONAL MID-RANGE BUILDING BLOCK COMPONENTS TO CONVERT THESE SPEAKERS TO 3-WAY SYSTEMS, AND JBL OFFERS A WIDE RANGE OF ADDITIONAL EXTENDED RANGE, LOW FREQUENCY, MID-RANGE, AND HIGH FREQUENCY DRIVERS PLUS PASSIVE RADIATORS AND CROSSOVER NETWORKS. NEITHER DOES THIS LIST INCLUDE THE E-V MC 8A AND JBL LE 8T USED IN PROJECTS 6 AND 7.

**Edlie Electronics, Inc.
2700 Hempstead Turnpike
Levittown, Long Island, NY 11756**

Carries a line of raw speakers and other electronic components. Catalog: $1 from above address.

ELECTRONICS CHAIN STORES

Lafayette
House brand speakers and components. Catalog available from stores.

Radio Shack
House brand speakers and components. Catalog available from stores.

Index

A
A box for your speaker 100
Alignment data, Thiele's 170
Allow for voice coil inductance 48
Arrays
  enclosures for 218
  mid-range 220, 297

B
Baffle, flat 239
Balance, make a simple 60
Bookend speakers 225

C
Cabinet, finishing your 121
Calculator method, Keele's 171
Ceramic tile
  enclosures 226
  speaker system 227, 304
Characteristics
  desirable enclosure 103
  of loudspeakers 9
Choke coil, inductance 94
Classic bass reflex 191
Closed box
  enclosures 125
  systems 141
Cloth, grille 121
Column, small PA 301
Compact
  closed box speaker system 155, 278
  ported box speaker 188
  ported speaker 287
  stuffed box speaker 230, 310
Compatability, woofer-tweeter 28
Compliance, increase 258
Component values 56
Cones
  conventional drivers with plastic 239
  increase mass of 261
  mass & compliance test 73
  resistance 19
  Vd, peak displacement volume of 86
Construction techniques & tools 107
Contemporary trends in speaker systems 235
Conventional drivers with plastic cones 239
Coupler & lead resistance 99
Critical listening 98
Crossover
  design a simple 6dB/octave 2-way 47
  frequencies to match drivers 33
  network 34, 42
  network, design 12dB/octave 52
  network, 3-way 12dB/octave 53
Curve, impedance 59

D
Damaged woofer 263
Damping
  material 118
  speaker 98
Design
  a passive radiator reflex 182
  a ported box for a driver 159
  a simple 6 dB/octave 2-way crossover 47
  a 6 dB/octave 3-way parallel network 50
  12 dB/octave crossover network 52
Designing an equalized reflex system 181
Desirable enclosure charactistics 103
Dispersion 14
Distortion & power handling 84
Drivers
  crossover frequencies to match 33
  design a ported box for a mid-range 159
  new kinds of 30
  single vs. multiple 236
  Walsh wave transmission line 26
   237

E
Efficiency 81
Efficiency, speaker 16
Electro-voice MC8A ported system 184, 281
Electronics chain stores 334
Enclosures
  ceramic tile 226
  closed box 125
  for arrays 218
  improve a cheap commercial 123
  materials 104
  new kinds of 239
  ported box 128
  shape affects sound 111
  types 125
  unusual speaker 224
Equalized reflex systems 178
Extension, microphone 62

F
F_b tuned frequency of ported box 78
Fibonacci series 113
Finishing your cabinet 121
Fit box volume to speakers 146
Flat baffle 239
Free air resonance test 63
Frequency response 11, 82

G
Golden ratio 114
Grille cloth 121
Guides, transmission line 204

H
Heil air motion transformer 237
Homemade
  test equipment 59
  tester for speaker polarity 88
Hook up extension speakers 248
Horns 136
How to test speakers & speaker system components 57
Hybrids 137

I
Impedance
  and phase method 80
  curve 59
  method 80
  speaker 20
  test 67
Improve a cheap commercial enclosure 123
Improving woofers 258
Increase
  compliance 258
  mass of cone 261
Inductance
  allow for voice coil 48
  of choke coil 94
  voice coil 92
Installation, speakers 114
Instruments, test 57

J
JBL ported system 186, 284

K
Keele's
  calculator equation for vented box frequency response 175
  method of computing vent dimensions 177
  pocket calculator method 171
Kinds, omnidirectional speakers 207

L
Labyrinth
  quarter wavelength 198
  tuning 198
Level & tone control settings 245
Line arrays
  for PA systems 218
  rock concert 220
Listening
  method 80
  tests 84
  test for stereo 91
Loudspeakers
  characteristics of 9
  enclosures, use pocket calculator to design 171
Low cost
  sub woofer 234, 312
  woofer-tweeter system 263, 313

335

M

Magnets	17
Mail order houses that offer speakers & components	332
Make a simple balance	60
Make a standard box	60
Mass & compliance method	75
Matching inductance to capacitance	95
Material	
damping	118
enclosure	104
Method 1	
precision resistor standard	67
VTVM	64
Method 2	
oscilloscope	64
variable resistance	69
Microphone	
extension	62
oscilloscope without	84
Mid-range	
array	220, 297
driver	30
Miscellaneous unconventional speaker systems	241
Mounting pattern	214
Multiple	
driver speaker systems	26
speaker arrays	213

N

Nearfield microphone technique	83
Network, crossover	34, 42
Newspaper test for speakers	88
Notes on using your speakers	243

O

Omnidirectional	
speakers	206
speaker system	208, 291
Oscilloscope	
test	88, 92
without microphone	84

P

PA column, small	223
Patching damaged speakers	262
Peak displacement volume of cone Vd	86
Phase corrected systems	240
Phasing, speaker	91
Piezoelectric tweeters	37, 236
Placement on baffle, speaker	116
Pocket calculator to design loudspeaker enclosures	171
Polarity	88
Polarity/damper tester	61
Ported box	78
enclosures	128
F_0 tuned frequency of tune a	78
tune a	167
Position, speaker	244
Power	
for extension speakers	250
rating	18
Project plans & construction notes	265
Protecting your speakers	246

Q

Q modification test	71
Q, speaker	22
Quarter wavelength labyrinth	198

R

Rainy day projects	253
Rating, power	18
Reflex	
classic bass	191
system	157
Replacement, speaker	243
Resistance	
coupler & load	99
cone	19
Response	
frequency	11
transient	25
Rock concert line arrays	220
RTR direct drive electrostatic tweeter	238
Rx for tweeters	256

S

Settings, level & tone control	245
Simple 2-way system	48, 267
Single	
column stereo system	221, 300
vs. multiple drivers	26
woofer stereo	232
Small PA column	223, 301
Speakers	
bookend	225
compact ported	287
compact ported box	188
compact stuffed box	230, 310
damping	98
efficiency	16
fit box volume to	146
for quad system	252
hook up extension	248
impedance	20
installation	114
newspaper test for	88
notes on using your	243
omnidirectional	206
patching damaged	262
phasing	91
placement on baffle	116
polarity, homemade tester for	88
position	244
power for extension	250
protecting your	246
Q	22
Q test	69
replacement	243
test	224
unconventional plastic diaphragm	238
utility bass reflex	195, 289
works	10
Standard test box	77
Stereo	
listening test for	91
single woofer	232
Stuffed boxes	228
System	
ceramic tile speaker	227, 304
closed box	141
compact closed box speaker	155, 279
contemporary trends in speaker	235
designing an equalized reflex	181
electro voice MC8A ported	184, 281
equazied reflex	178
JBL ported	186, 284
line arrays for PA	218
low cost woofer-tweeter	263, 313
miscellaneous unconventional speaker	241
multiple driver speaker	26
omni speaker	208, 291
phase corrected	240
reflex	
simple 2-way	48, 267
single column stereo	221, 300
speakers for quad	252
3-way closed box speaker	150, 272
3-way speaker	55, 269
2-way closed box	153
2-way closed box speaker	276

T

Tame bad peaks	253
Test	
box, standard	77
cone mass & compliance	73
equipment, homemade	59
free air resonance	63
impedance	67
instruments	57
listening	84
oscilloscope	88
procedures	62
speaker	224
Q modification	71
Q speaker	69
V_{as}	74
white noise	86
Thiele's alignment data	170
3-way	
closed box speaker system	150, 272
speaker system	55, 269
12 dB/octave crossover network	53
Tools & construction techniques	107
Towers	241
Transformer, Heil air motion	237
Transient response	25
Transmission lines	199
labyrinths	133
guides	204
Tune a labyrinth	198
Tune a ported box	167
Tweeters	
piezoelectric	37, 236
RTR direct drive electrostatic	238
Rx for	256
2-way closed box speaker system	153, 276
Types of enclosures	125

U

Unconventional plastic diaphragm speakers	238
Unusual speaker enclosures	224
Utility bass reflex speaker	195, 289

V

V_{as} test	74
Vent dimensions, Keele's method of computing	177
Visual method	80
Voice coil inductance	92
VTVM test	92

W

Walsh wave transmission line driver	237
White noise test	86
Wiring circuits	214
Woofer	
damaged	263
improving	258
low cost sub	234, 312
tweeter compatibility	28